Library Collection Assessment through Statistical Sampling

Brian J. Baird

The Scarecrow Press, Inc.
Lanham, Maryland • Toronto • Oxford
2004

SCARECROW PRESS, INC.

Published in the United States of America
by Scarecrow Press, Inc.
A wholly owned subsidiary of
The Rowman & Littlefield Publishing Group, Inc.
4501 Forbes Boulevard, Suite 200, Lanham, Maryland 20706
www.scarecrowpress.com

PO Box 317
Oxford
OX2 9RU, UK

British Library Cataloguing in Publication Information Available

Library of Congress Cataloging-in-Publication Data

Baird, Brian J., 1966–
Library collection assessment through statistical sampling / Brian J. Baird.
p. cm.
Includes bibliographical references and index.
ISBN 0-8108-5038-9 (pbk. : alk. paper)
1. Collection management (Libraries)—Statistical methods. 2. Sampling (Statistics) 3. Library use studies. I. Title.
Z687.B35 2004
025.2'1'0727—dc22

2004006774

™ The paper used in this publication meets the minimum requirements of American National Standard for Information Sciences—Permanence of Paper for Printed Library Materials, ANSI/NISO Z39.48-1992.
Manufactured in the United States of America.

To William J. Crowe, who has done so much
to help me throughout my career,
and to my friend Bradley L. Schaffner,
who has worked closely with me
on many collection assessment projects.

Contents

Figures and Tables

Figures

Tables

Acknowledgments

I want to thank my wife, Jennifer, and our seven children for their continued support of my professional endeavors. I also want to thank the good folks at Scarecrow Press who have been so helpful in bringing this book to press—especially Dr. Sue Easun who has been patient and encouraging with the preparation of this book and with other publishing projects I have brought to her attention.

Introduction

Assessment is a hot topic in libraries. It is recognized that the evaluative information gained through well-performed assessment can be used to improve a library's collections, services, physical layout of the building, and so forth. However, because assessment has become so popular, libraries often find themselves feeling pressure to evaluate themselves without really knowing specifically what needs to be evaluated. Assessment should always be designed to provide information about the collections or services you provide in order to direct how well your library is fulfilling its mission. The mission guides the assessment, and the assessment results help shape future goals. Assessment results help keep goals outcome oriented. Otherwise, library assessment is like the enthusiastic sailor who decides to build his own sailboat. He plans on a design, gets the materials, builds a great boat, only to discover after the boat is finished that it is too big to fit out the door of the garage.

Assessment is a planning tool, but the assessment itself must also be carefully planned, shaped by the mission of the library, to ensure that the assessment exercise is effective and provides valid, useful information. There are many very effective library assessment tools available, and these tools can be used to gain useful information, but often these off-the-shelf assessment tools are so complex and so involved that they require too much time and effort from the library staff and patrons to complete. That is why the assessment principles, strategies, and tools discussed in this book help guide a library down a decision-making path that will result in a well-designed assessment tool that is easy to use and will provide the library with the information they need and no more.

When planning to do an assessment, it becomes extremely tempting to gather too much information. The reasons for this are discussed in the first chapter, but it is important to understand now that this book has been designed to provide "how to" strategies for assessing library collections based on principles of efficiency and economy of effort.

Throughout this book librarians will be asked to answer the following three questions:

- What do you want to evaluate?
- What information is needed to make that evaluation?
- What will it take to get that information?

Each of these questions beg other questions, but these are the primary questions that must be answered before an assessment program is begun, and the answers must be referred to regularly to keep the assessment project on track.

There are lots of assessment strategies and tools available such as surveys, focus groups, user statistics, and so forth. The focus of this book is to provide detailed instructions on how to conduct an effective assessment program using statistical sampling of a library's collections.

The first chapter of this book provides the philosophical groundwork about how to conduct a collection assessment. This information is useful for better understanding how to utilize the information in the subsequent chapters that provide detailed instructions on how to effectively assess your library's collections.

Chapter 1

Purpose of Assessment

Assessment is a tricky business. We often feel we do not have enough understanding to know what information we need. This frustration can cause us to overdo our efforts to gain evaluative information about the services, collections, and practices of the library. A very common fear with assessment is that a great deal of effort will be put into gathering and analyzing information only to find that the information is incomplete. This is why it is important to use proven assessment techniques or hire experienced professionals to complete the assessment. Hiring is the easiest process, but it is also expensive—often restrictively so. Many of the assessment tools that have been developed are robust and provide a wealth of data, but again, often the tool does not adequately fit your institution.

The techniques and strategies described in this book are designed to provide sound assessment principles from which a library can customize their own assessment tools to gather the information they need. Most of the examples used in this book relate to library preservation and collection development, but the techniques can easily be adapted for other library services once you understand the principles behind the assessment strategies presented here.

Most often we want assessment information about the library from a number of different viewpoints. We want to not only know how our collections stack up against other peer libraries, but we also want to know if our patrons feel that our collections have the information resources that they need to further their research. For this reason, I very much like the principles presented in the Balanced Scorecard method of assessment.

The Balanced Scorecard[1] method of assessment was developed by Robert S. Kaplan and David P. Norton in the early 1990s and has been used successfully in a wide array of business settings. Based on the premise that "what you measure is what you get,"[2] the Balanced Scorecard "is like the dials in an airplane cockpit: it gives managers complex information at a glance."[3] In other words, the design of the Balance Scorecard method is to provide big picture assessment of a business or organization by giving a "balanced presentation of both financial and operational measures."[4] In recent years the Balanced Scorecard has begun to be used as an assessment tool in academic settings[5] and research libraries.[6]

The Balanced Scorecard appeals to the assessment needs in libraries because it represents a "balance between external measures for shareholders and customers, and internal measures of critical business processes, innovation, and learning and growth."[7] Libraries are driven by the needs of the users, collection development policies and practices, available funding, and the flow of materials through processing units in the library. Therefore, when conducting an assessment program, it is important to take a wide and thorough look at the library as a whole and not just at one aspect. For this, the Balanced Scorecard method provides a very effective tool.

Libraries are becoming increasingly interested in the services they are providing for their users. This is an important focus—especially as more and more information becomes available electronically. However, the traditional strengths of libraries has always been their collections. This is true still today—especially in research libraries. Also, collection makeup is the hardest thing to change quickly. For example, if a library has a long tradition of heavily collecting materials published in Mexico, then even if that library stops purchasing all Mexican imprints, its Mexican collection will still be large and impressive for several years to come unless they start withdrawing books. Likewise, if a library has not collected much in a subject, and then decides to start collecting heavily in that area it will take several years for the collection to be large enough and rich enough to be considered an important research tool.

For this reason, assessment efforts in a library need to evaluate the makeup and use of the collections it owns. The collections are the foundation upon which other services and activities are based, thus the assessment strategies outlined in this book focus on assessing a library's collections.

Building strong library collections is often a difficult challenge due to the constant obstacles of limited funding, insufficient staffing, and an unending demand on library resources. But librarianship is about managing resources.

Library preservation makes a good case study for other library assessment activities in that at the end of the day, after the library is staffed; materials acquired, processed, and placed on the shelves; and the light bill paid, there is little funding left for preservation or much else. Preservation is always considered important to library professionals, but other than dealing with disaster recovery efforts, preservation does not command urgency. It often gets slotted to the back burner and remains there until someone decides to champion it. This is like so many other library endeavors where urgent remedial endeavors crowd out the important but nonurgent strategic endeavors the library would like to focus on. This is why so many assessment results and strategic planning efforts do not produce major changes in a library. The daily activities of maintaining a library keep the staff and administration focused on the here and now, leaving no time and energy to focus on the future. The only way to change this fact is for the library's administration to encourage assessment and planning efforts. They have to work hard to get people to focus on the future until it becomes part of the culture.

One way to help this happen is for the administration to recognize, and make, the easy changes that can be made. This will give the library staff a sense of confidence that change can happen and that it can be positive. Let's look at preservation as an example for this. Again, librarians generally feel that preservation is important, but it is often difficult to commit the resources necessary to provide a library with the level of preservation it deserves. Therefore, to implement an effective preservation program, a library must focus on preservation strategies designed to reduce immediate and long-term costs and strengthen the library's collection development goals.

The first step when beginning a preservation program is to clearly document the preservation needs of the collection. Again, focus on the foundation of the library—its collections. A proper assessment involves establishing usage patterns for the materials in the library, thoroughly evaluating the collection development plan, assessing the potential threats to the collection, and identifying the preservation resources readily available. The Balanced Scorecard is an ideal tool for collecting this information. Each library is different, meaning assessment strategies will also have to be different, but the following assessment tool can generally be used to develop a strong preservation plan for a library of any size.

By using the Balanced Scorecard assessment approach, a library can evaluate all operations, both public and technical. This provides institutional context and keeps the assessment effort grounded in reality. The Balanced Scorecard approach can help a library develop a preservation strategy created to best meet its specific needs. This is because an assessment approach will allow a library to identify its preservation concerns and challenges, thus providing a clear indication of what the library should do, over time, to improve its preservation efforts.

Preservation efforts are most effective when they are based on the practical needs of the library. Few libraries can afford to have an ideal preservation program that fully addresses the needs of its collections. Therefore, the following Balanced Scorecard assessment tool is designed to help a library identify areas of concern, optimize the use of their resources put toward preservation, and establish goals for continual improvement over time as additional resources become available.

The Balanced Scorecard links performance measures in four areas: external perspective (services provided); internal perspective (effectiveness and efficiency of daily operations); financial (resources); and innovation (future planning and direction). More specifically, figure 1.1 shows how a Balanced Scorecard assessment in a library would look for preservation planning:

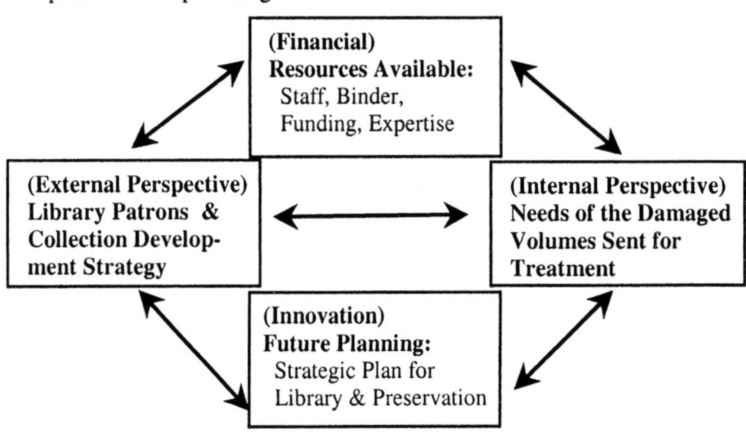

Figure 1.1: Balanced Scorecard for Preservation Planning

By looking at each of these performance measures individually, it becomes clearer how they impact each other.

Needs of the Damaged Volumes Sent for Treatment

We begin with this performance measure because for a preservation program this is where the tire meets the road. Treatment decision making is the first step undertaken on any item sent for repair or binding. However, it is difficult to discuss strategies for how best to treat a damaged item without clearly identifying the needs of the collections and the resources available. Again, this is why the Balanced Scorecard approach is so effective in assessing a library's preservation needs.

Most preservation treatment programs are driven by use, meaning most damaged items sent to a preservation unit for repair are identified as needing treatment when they return from circulation. It is, therefore, important to understand the makeup of the materials circulating in the library as well as the makeup of the collections in general. This is best done through a careful collections condition survey.[8] As will be seen, an effective collections condition survey, planned from a Balanced Scorecard perspective is an incredibly powerful preservation assessment tool. Likewise, it is important for a library to understand how the collections they own are being used by patrons. This will help guide future collection development activities.

Each library must create its own survey instrument to meet the specific needs of their collections because each library has strengths and weaknesses that will impact how they conduct their survey. Another preservation example is some libraries are fortunate to have skilled staff to provide book repair or conservation treatment. Other libraries find themselves served by a good library binder who can meet most of their binding and repair needs. If a library has an active and developed book repair or conservation program the survey will need to be structured to gather specific information on how many items need in-house repair or full conservation treatment. This information will help the library plan staff time, supply budgets, and ways to improve in-house treatments (if survey results uncover inherent weaknesses in some of the treatments being performed).

A collection development example might be a library that has run out of growth space and is contemplating a weeding project. Knowing this would shape the kind of collection survey that a library conducted.

The first thing to decide is what information a library needs to gather. This decision will effect how the survey is conducted. General information can be gathered using a relatively small random sample. Detailed information about the collection will require a larger sample in order for the results to be statistically valid.

In addition to information gathered about the entire collection, it is also important to assess the types of materials being used by patrons. Many libraries—especially older libraries—will find that the data about the volumes they circulate to patrons is quite different from the makeup of their general collections. Therefore, it is important to sample the population you want the data about and to not over generalize. For example, a large old research library might have a collection in which 25 percent of the holdings have brittle paper. However, it may be found that only 5 or 6 percent of the materials that circulate are brittle. When developing preservation strategies for dealing with brittle books, it is important to understand this information. If your library is only going to treat volumes that circulate, then it only needs to establish a brittle book fund for 5 percent of the volumes that circulate.

Treatment decisions involve many levels. There is always an economic aspect to every decision. For example, is it more cost effective for your library to discard the damaged item in hand and replace it with a new copy? Is it less costly to bind the item, or have it repaired in-house? Experience and practice will help make these decisions easier, but the dilemmas will never completely disappear. Again, an effective collection assessment can be a very useful tool to help answer treatment decision-making questions.

A strategy for making treatment decisions must be based on the needs of the collection and the resources available—including funding, vendors, and staffing. It must also take into account the collection development policies of the library. For this reason, the treatment decisions will be unique in each individual library. The same is true for other library endeavors whether it be circulation policies, library hours, and so forth. Each library must know who they serve, what services are required, and how best to supply those services. Answering these questions requires understanding of all four Balanced Scorecard performance measures of: external perspective, internal perspective, financial, and innovation.

Many library decisions are best understood when viewed as a continuum. For example, some collection development continuums are electronic resources versus print collections; reference collections versus circulating volumes; or research collections versus teaching support collections. There are only so many collection funds per year, and choices have to be made between electronic resources and print collections, or having a strong reference collection or a stronger circulating collection.

Again, preservation serves as a good example of how many library decisions can be viewed as a continuum (see figure 1.2).

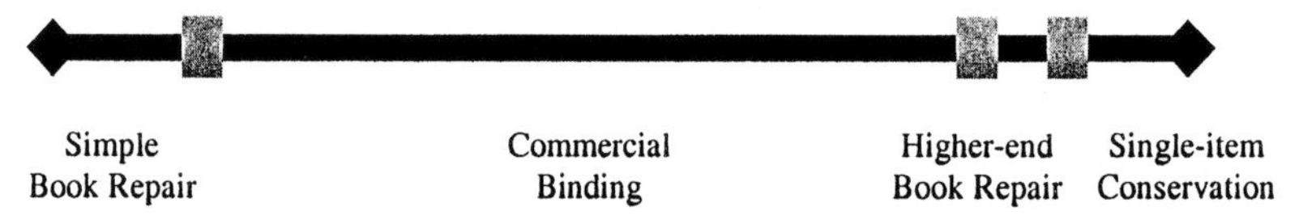

Figure 1.2: Treatment Decision-making Continuum

There are many simple repairs like tip-ins and simple paper mending that can easily be done in-house. The vast majority of materials in most libraries can best be handled by a commercial bindery. On the other end of the continuum is higher-end book repair and single-item conservation treatment.

It does not take very much experience in preservation treatment decision making before it becomes clear why it is best viewed as a continuum. Many decisions are easy, but things become less clear as they approach the line between the various treatment options. For example, book repair can fall into either end of the continuum depending on the resources of the library. If a library has a qualified staff member who can perform various kinds of book repair in an efficient manner, and if that library cannot obtain inexpensive commercial binding, it may be less expensive for that library to perform as many book repairs as possible rather than doing a lot of rebinding. For such a library, more book repairs will fall into the simple book repair part of the continuum, and the amount of area on the continuum assigned to commercial binding will be smaller. By contrast, a library that has limited staff resources and a good binding contract may find it is cost effective to rarely perform book repairs in-house and to instead send everything to their library binder. For this library, book repairs would mostly be higher end book repair treatments reserved for those items that need to be rush repaired, or have special needs that require it to be treated in-house rather than at a bindery.

Only through careful assessment of the resources available to the library and its collection development policy can each library decide where the lines fall for them on their various decision-making continuums. Every library, regardless of size, faces the same challenge when grappling with a decision-making policy. Every line on the treatment decision-making continuum represents a gray area where the tough decisions need to be made. For this reason, the graphic in figure 1.2 is presented with thick gray lines to suggest the ambiguity involved with treatment decisions that fall close to the edges of each area on the continuum. These tough decisions are everywhere in library management. Careful and continuous assessment is the only way to have confidence that you are making good decisions that take into account the collection development policy for the library, the needs of the collection, and the resources available both internally (staff and funding) and externally (library vendors). In other words, decision-making is always an evolving, learning process that is improved upon through experience and evaluation of the process.

Library Patrons and Collection Development Strategy

The collection development needs of a library cannot be determined outside of the context of its patrons' usage patterns and the library's mission and strategic plan. Conducting collection development activities outside of this context will result in ineffectively used resources and frustration. For example, if the patrons' informational needs are not met, the library will receive constant complaints, or the patrons will stop coming to the library. It can sometimes be difficult to determine what the patrons want. This is because they often do not know what informational needs they will have until they are presented with those needs.

Too often patron surveys ask leading or non-qualifying questions that do not produce good results. For example, I have seen surveys that will ask, "Are you willing to wait five to ten minutes to use a computer when you come to the library?" It should not be surprising to anyone to find out that patrons are not willing to wait five to ten minutes to use a computer, but that information does nothing to tell the library how many computers it needs. So while user surveys can be beneficial, they also have to be supported by data on how the library's collections and infrastructure are being used. A survey of the library's collections can determine usage patterns in the collection as well as the preservation needs of a collection. It provides useful data that will directly impact the collection development plan. For example, the results of a collection assessment can yield findings on the kinds of physical formats that receive the most use, and what subject areas are most heavily trafficked. The survey results can help guide the collection development efforts to strengthen those areas that receive the heaviest patron use.

There are many questions to ask during the book selection process. For example, there are many titles that can be purchased simultaneously in either publisher hard-cover bindings, publisher paperback bindings, or vendor-prepared bindings designed to hold up to library use. Which choice is best depends on the title itself. If a title does not circulate very often, dollars spent on sturdier bindings are not well used. For this reason it is important for a library to know its collections' usage patterns. Book selection practices that take into account the cost effectiveness of various binding formats will eventually translate into significant savings for the library, which will lead to better services and more materials for the patrons.

Resources Available

After the makeup and usage patterns of the collection are determined, this information can be used to direct collection development practices, but these decisions must also be based on knowing what resources are available. Again, the strength of the Balanced Scorecard approach is that it forces the library to look at more than one area of the library when conducting an assessment. Again, using preservation as an example, some libraries have staff members who are well trained at performing book repair techniques. Others have fully staffed and equipped conservation facilities. And many libraries have favorable contracts or agreements with their library binder. By balancing the preservation needs of the collections, and the resources available that can be assigned to the problems, a library can develop strategies to optimize its preservation efforts.

Many libraries are not fully aware of the preservation resources available to their library simply because they have never looked into it. For example, most libraries employ the services of a library binder, but many are not fully aware of all the services offered by their binder. There are also a surprising number of training opportunities available where library staff can get formal training in specific preservation activities. Just as it is important to assess the collection to determine its preservation needs, it is also necessary to carefully assess the preservation resources available to your library. The same is true for collection development activities. For example, with the increase of electronic resources available, there are many consortial arrangements that can be made to make it easier for libraries to acquire access to more information than they would be able to otherwise secure on their own. A library has to remain continually vigilant on identifying new opportunities for securing access to new information sources—especially electronic packages.

Funding is usually the most difficult challenge a library faces for any endeavor it undertakes. An important part of any assessment effort is identifying what it will cost to implement various services in your library, and how funds can be made available to secure those services. When done properly, the assessment process will identify ways of saving money that can then be used to further other efforts in the library.

Future Planning

Future planning has been addressed several times in this discussion. To have an effective library—especially in this current electronic age—careful assessment and planning must be ongoing. However, there is also a need to establish some long-term goals based on the information gathered from the Balanced Scorecard assessment. The collection survey, the library's patron usage patterns, and the analysis of the resources available will combine to identify long-term preservation needs of the library.

For example, the collection survey may identify that the library has a high percentage of brittle materials that are being used regularly, necessitating the need to begin a reformatting program. There may not be funding sufficient to meet these reformatting needs, so long-term plans can include grant writing or fundraising opportunities. Another example is a library that discovers from the condition survey that it needs a full-time conservator to treat the high number of damaged materials needing single-item treatment. This means hiring a high-level professional, providing support staff for the position, and constructing a conservation laboratory. This is an expensive proposition that will not happen without a long-term commitment by the library.

Finally, the Balanced Scorecard assessment, combined with the library's strategic initiatives, will help the library identify areas that will need further evaluation, or future assessment, to measure if progress is being made. For example, all libraries are being bombarded by various issues surrounding electronic information. This rapidly changing climate requires closer monitoring than other areas of the library. The decisions made today will need to be assessed regularly in the future to ensure that strategies remain viable.

Practical Examples

It can be difficult to take theory and put it into practice. But looking at the preservation activities at the University of Kansas (KU) Libraries can demonstrate how the Balanced Scorecard approach can help to address all of the assessment and planning issues faced by a research library's preservation program.

The Balanced Scorecard approach looks at four performance areas (see figure 1.3).

(Financial) **Resources Available** Staff, Binders, Funding, Expertise	**(External Perspective)** **Library Patrons &** **Collection Development** **Strategy**
(Innovation) **Future Planning** Strategic Plan for Library & Preservation	**(Internal Perspective)** **Needs of the** **Damaged Volumes Sent** **for Treatment**

Figure 1.3: Balanced Scorecard Preservation Performance Measures

As stated earlier, one of the most effective assessment tools that can be used to help address these performance areas is a thorough collection condition survey. At the University of Kansas Libraries we have conducted several condition surveys ranging from our large survey[9] in which over 3,500 items were examined, to smaller surveys[10] focused on individual collections or surveys designed to gather very specific information about individual subject collections. From these surveys we have gained information that has shaped decisions made in each of the four performance areas. To illustrate this more we will look at each performance area individually.

Needs of the Damaged Volumes Sent for Treatment:
It is obvious how a condition survey can be useful in assessing this performance area. The very purpose of most condition surveys is to learn what percentage of a library's collection is damaged, and to gain information about that damage. But at KU we also learned how well previous treatments held up to use and what we could do to improve them. For example, we learned that commercially bound volumes rarely require future treatment,[11] and that the case bindings we produce in our conservation lab hold up extremely well to heavy use.[12]

Condition surveys can also identify preservation problems that need to be addressed. An example of this can be seen in my experience while working at Princeton University. An assessment of materials returning from circulation showed that a large number of eighteenth- and early nineteenth-century materials regularly circulated. Therefore, it was necessary to develop a specific treatment to address the preservation needs of these volumes in a production setting of a collections conservation program.[13] This was a problem fairly unique to Princeton University based on some research requirements each student had to fulfill in order to graduate. Therefore, the needs of the damaged volumes were dictated by patron usage patterns.

Library Patrons & Collection Development Strategy:

At KU, the condition survey results greatly impacted the collection development strategies for the libraries. We learned that paperback volumes hold up well in our stacks, so we switched to a paper preferred collection policy that saves thousands of dollars a year.[14] We also changed some of our binding strategies to take advantage of some economy binding options available from our binder.

Policies for when and how your library accepts gift collections can also have huge preservation implications. For example, if a large gift of brittle volumes is accepted by the library it can quickly overwhelm the preservation reformatting budget and staff-processing capabilities.

Resources Available:

A condition survey of materials returning from circulation can quickly identify the major preservation challenges facing a library's collections. This will enable the library to estimate what it will cost per year to repair the damaged items sent to the preservation program, and will quickly let a library know if it has supplied sufficient funding to its preservation program. However, in many ways, it is important to know what preservation resources are available before you conduct the survey so that the right kinds of data can be gathered during the survey process. For example, we knew that the KU Libraries had a quality binder who was offering our library some economy binding services that were not available to us from our previous binder. Therefore, we structured the questions of the survey to learn how we could best use these new services.[15] We also had a good understanding of what treatment options were available in our conservation laboratory and from our contracted vendors before we began the survey.

The condition survey can be very useful in gathering information that will help with fundraising and grant-writing efforts. The data gathered from the general condition survey and the survey conducted on the Slavic materials in the KU Libraries proved very useful in strengthening several grant applications. The information can also be useful in helping a library decide what kind of staffing the preservation department needs and how best to use student employees.[16] Knowing the needs of the collections helps the library better plan what mass, lower level treatments need to be performed such as pamphlet bindings, paperback stiffening, or dust jacket covers.[17]

Future Planning:

Research libraries have been engaging in strategic planning for many years. The preservation program should also develop long- and short-term goals of its own.[18] It is important for the preservation program's plans to compliment and support the library's overall goals. For example, the collection survey at KU demonstrated that while our brittle books problem was not as severe as those at other large research libraries, we still had far more brittle books circulating and needing treatment than we could accommodate at our current funding level. Therefore, we introduced some cooperative efforts[19] to reduce the cost of preservation photocopying, and we began long-term fundraising and grant writing plans to subsidize our reformatting budget.

An important part of future planning is continual assessment. The initial condition survey will provide a wealth of information, but it will also stir up a large number of questions that will need further study. Also, the changing nature of library collections and how we provide access to information for our patrons will impact how collections are used, and will require careful monitoring. In short, effective planning for the future necessitates the need for ongoing assessment.

Based on this discussion we can look again at our table of performance areas, and can add specific questions that need to be assessed for each area (see figure 1.4).

Resources Available	**Collection Development Strategy for Library**
• Is funding sufficient to meet preservation needs of collections? (i.e., binding, reformatting, and supplies budgets) • Does the preservation program have sufficient staffing? • Does the preservation program have the right kind of staff? (i.e., student employees, support staff, professional staff) • Does the library need additional preservation expertise? (i.e., preservation administrator, conservator, reformatting specialist) • Has the library investigated outside funding options for preservation? (i.e., grant writing, fundraising)	• Should the library have a paper preferred policy? • What is the library's prebinding policy? • What are the preservation implications for accepting a gift collection of books? (i.e., brittle volumes, volumes needing repair) • What are the preservation implications for the various collecting strengths of the library? (i.e., non-western collections are printed on poor-quality, acid paper with weak bindings) • Does the library need to engage in a weeding project of some kind—especially to remove multiple copies

Figure 1.4: Expanded Balanced Scorecard Preservation Performance Measures

Future Planning	Needs of the Damaged Volumes Sent for Treatment
• What are the preservation implications of the library's strategic plan? • How can the preservation program help the library meet its goals? • What are the long-term plans for the preservation department? • What obstacles need to be overcome for the preservation program to meet its goals? (i.e., space issues, funding, staffing, training) • What additional assessment is needed? • What impact will digital access have on preservation activities?	• What percentage of the collection is damaged enough to require treatment or rebinding? • What percentage of the collection is brittle? • How effective are the current repairs? • What new preservation initiatives need to be started to treat the damaged collections? • How will digital information impact the future usage needs of paper based information?

Figure 1.5—Continued

Conclusion

The Balanced Scorecard is an effective assessment tool for use in a library. This assessment strategy is particularly effective when linked with a thorough collection survey. Many organizations that have used the Balanced Scorecard have found that the effort not only resulted in producing the assessment information they needed at the time, but also revealed "an opportunity to use the Balanced Scorecard in a far more pervasive and comprehensive manner than originally intended."[20]

The Balanced Scorecard can be a very effective tool for assessing the needs of a library in other areas as well. It could also be used to assess how effective the library is in fulfilling its institutional mission. It is simply a matter of identifying what needs to be assessed and how best to assess it in the four general areas of services provided, effectiveness and efficiency of daily operations, financial, and future planning and direction.

Notes

1. Robert S. Kaplan and David P. Norton, *Translating Strategy into Action: The Balanced Scorecard* (Boston: Harvard Business School Press, 1996).
2. Robert S. Kaplan and David P. Norton, "The Balanced Scorecard—Measures That Drive Performance," *Harvard Business Review* 70 (January-February 1992): 71.
3. Kaplan and Norton, "The Balanced Scorecard," 71.

4. Kaplan and Norton, "The Balanced Scorecard," 71.
5. See for example Alice C. Stewart and Julie Carpenter-Hubin, "The Balanced Scorecard: Beyond Reports and Rankings," *Planning for Higher Education* 29 (No. 2, 2001): 37-42; and Harold F. O'Neil Jr. et al. "Designing and Implementing an Academic Scorecard," *Change* 31 (Nov./Dec. 1999): 32-40.
6. Jessica Clark, "Balancing the Scorecard of User Experience," Bibliozine.com: The E-magazine for Librarians 1 (January-February 2001) at: http://www.bibliozine.com/articles/012001/04.shtml. The Balanced Scorecard method has been popular in research libraries in Australia. To learn more, see the Deakin University Library report on their implementation of the Balance Scorecard on their web site at: http://www .deakin.edu.au/library/staff/LIANZABSCarticle.htm.
7. Kaplan and Norton, *Translating Strategy into Action,* 10.
8. Brian J. Baird, Jana Krentz, and Brad Schaffner, "Findings from the Condition Surveys Conducted by the University of Kansas Libraries," *College & Research Libraries* 58 (March 1997): 115-126.
9. Baird, Krentz, and Schaffner, "Findings from the Condition Survey."
10. Bradley L. Schaffner and Brian J. Baird, "Into the Dustbin of History? The Evaluation and Preservation of Slavic Materials." *College & Research Libraries* 60 (No. 2, March 1999): 144-151.
11. Brian J. Baird, "Those Amazing Library Bindings," *New Library Scene* 15 (October 1996): 9-10.
12. Brian J. Baird, "Case Binding with Flexible Bonnet: A Specification for General Library Collections," *New Library Scene* 13 (October 1994): 8-10.
13. Brian J. Baird and Mick LeTourneaux, "Treatment 305: A Collections Conservation Approach to Rebinding Laced-on-Board Binding Structures," *Book and Paper Group Annual* 13 (1994): 1-4. Also on the World Wide Web at: http://aic.stanford.edu/conspec/bpg/annual/v13 /bp13-01.html.
14. Brian J. Baird, "Paperbacks vs. Hardbacks: Answers from the University of Kansas Libraries' Condition Survey," *Abbey Newsletter* 20 No. 6 (December 15, 1996): 93-95. Also on the World Wide Web at: http://palimpsest.stanford.edu/byorg/abbey/an/an20/an20-7/an20-708.html.
15. Baird, "Those Amazing Library Bindings."
16. Brian J. Baird, "Motivating Student Employees: Examples from Collections Conservation," *Library Resources & Technical Services* 39 (1995): 410-416.
17. Brian J. Baird, *Preservation Strategies for Small Academic and Public Libraries.* (Lanham, MD: Scarecrow Press, 2003): 7-28.

18. Brian J. Baird, "Goals and Objectives of Collections Conservation," *Restaurator* 13 (1992): 149-161.
19. Brian J. Baird, "Brittle: Replacing Embrittled Titles Cooperatively," *College & Research Libraries News* 58 No. 2 (February 1997): 83-84, 95.
20. Kaplan and Norton, *Translating Strategy into Action,* 291.

Chapter 2

Designing the Survey Instrument

Before an effective strategy can be established for collection development or preserving a library's collections it is important to determine what materials a library holds, how they get used, and who are the primary users. These questions can be answered through a carefully conducted statistical survey of the collections. There are excellent publications that explain statistics and random sampling. Often the best place to start is with an introductory statistics textbook for social science or political science students. These applied types of textbooks explain statistical principles in a way understandable to the non-mathematician. They also give many real world examples of how statistics and sampling strategies can be applied. This background information can be helpful in conducting various surveys to obtain specific kinds information—whether it is usage patterns, condition of the collections, or patron information needs. There are also many service bureaus and consultants throughout the country who will contract to conduct library surveys if your library has the money to pay for these services.

The survey instrument described in this chapter is designed for libraries to pick and choose from the questions provided to produce a collection survey that will provide the library with the specific information it needs. The underlying premise of this tool is to get exactly the information a library needs. Assessment is good, but a library can over-assess. Evaluative information is powerful, but sometimes it can cost more in time and money than it is worth.

Surveys are important, but librarians must also apply their rich experience to determine the needs of their collections and their users. Careful, long-term, documented observation about collection use, damage to materials, and types of materials in the collections can be invaluable in determining the strategic needs of a library. This conventional wisdom should be validated through assessment efforts, but too often assessment efforts start from ground zero and do not take advantage of the institutional knowledge and experience that exists.

As part of a preassessment effort, the library should document what they know about their library including the makeup of its collections, the strengths of its staff, its primary users, and so forth. The library needs to answer the following questions:

- What kind of a library is it? (e.g., public library, research library, college library, corporate library.)
- Who are the patrons and how do they primarily use your library?
- How are materials selected for the library's collections? (e.g., approval plans, bibliographer selection.)
- What parts of the collections are most heavily used and why?
- Does your library process materials for the shelves in-house, or is the work outsourced?
- Does your library pre-bind books or provide other shelf preparation treatments such as stiffening paperbacks or reinforcing dust jackets?

Answering these broad questions will help a library better select questions for surveying their collections. How your library answers the questions above will help you create a survey that will answer the following questions:

- Do prebound or stiffened covers hold up better than nontreated books? (This is an important question to answer. If they do not hold up significantly better, then the resources put into prebinding may be better spent in other areas.)
- Can you identify a pattern in how or where repaired books fail? If so, what can be done to address this inherent weakness?
- How many copies of a popular title are enough?

The answers to these questions, and others, will be different for each library, but there are some general guidelines in this chapter that will help libraries think about their collections in new ways.

Gathering Data

There are many ways to gather statistically valid information that will help you in identifying the makeup and needs of your collections. Some involve selecting random samples of books and examining those books in a consistent way. Such collection surveys can provide a wealth of information in a relatively short amount of time. However, many libraries routinely collect lots of information that can be useful in assessing the needs of their collections. For example, circulation records can be loaded into a database program that can identify what kinds of books circulate, how many times, on average, materials are checked out, what age groups borrow the most materials, and so forth. Some library systems make gathering such data easier than others, but nearly every system makes it possible to do at least some such analysis. User groups for your integrated library system can be a good source for finding pre-made queries for pulling user information.

Another simple way of identifying how materials get used and how they fail is to gather data on items that have been set aside for repair, rebinding, or replacement. Over a period of time extending several months, record the following information about each damaged book that is pulled from circulation. Identify:

- What kind of book is it, (e.g., popular novel, children's book, "how to" book)?
- What kind of binding does the book have, (e.g., paperback, hardbound, prebound)?
- How many times did the items circulate?
- What kind of damage caused the book to be pulled from circulation?

Using this data a library can identify what kinds of books are failing and why. Such information is invaluable for helping the library make informed purchasing decisions about what kind of materials to buy for the library, what kind of bindings these materials should have, and what kind of preshelving treatments are most effective. Such information can also be useful for determining which in-house repair techniques are effective and which are not.

In addition to collecting information about materials pulled from circulation because of damage, libraries can survey their stacks to learn a great deal about the physical condition of the materials in their collections. A randomly selected sample of only a few hundred books can provide a wealth of general information about the collection, provided the survey is well prepared and conducted in a statistically valid manner. This is the primary surveying tool described in this book. All other assessment data should be used to compliment the data gathered from the collection assessment.

Following is a list of survey questions a library can select from when surveying their collections. A library may wish to use all of these questions, but it is probably not necessary to do so. Simply select the questions that will be most relevant to your library. Remember, the temptation is to overassess. That is why it is important to know what information you *really* want before you create your survey instrument so that you do not waste time gathering data you think will be interesting, but in the end will not really be useful in helping to shape the strategic future of your library. Likewise, the scripted answers can also be altered for each question. For example, if your library does not have a conservation program you probably will not need as much detailed information about the types of damage that occur. Likewise, if your library does not support a strong international collection or area studies collection, you may not need as much information about where books were published. For this reason, on some of the questions there is a list of alternate scripted answers.

Collection Survey Questions

Size of Volume
- Regular
- Folio
- Oversized

Type of Volume
- Monograph
- Part of Multi-volume Set
- Serial
- Music Score

Leaf Attachment
- Sewn Through the Fold
- Oversewn
- Adhesive Bound
- Stapled Through the Fold
- Side Sewn or Stapled
- Spiral or Other Loose Sheet Binding

Condition of Text Block (Mark all that Apply)
- In Good Condition
- Remain in Stacks
- Needs Treatment
- Must be Replaced or Withdrawn
- Broken or Loose Sewing or Adhesive
- Broken Text Block
- Loose Pages
- Damaged Pages (not mutilation)
- Missing Pages (not mutilation)
- Pages Damaged or Curled from Lack of Support (Paperbacks)
- Multiple Pages Torn near the Gutter (Mostly Children's Books)
- Mold Damage

(Alternate list if less detail is needed)

In Good Condition
Remain in Stacks
Needs Treatment

Gutter Margin Width
- Less than ½ Inch
- More than ½ Inch, but Less than ¾ Inch
- More than ¾ Inch, but Less than 1 Inch
- 1 Inch or More

Paper pH
- Yellow or Clear (Acidic)
- Tan (Slightly Acidic)
- Purple (Alkaline)

Paper Fold Test (Paper Breaks After)
- Less than 1 Fold
- Less than 1 Double-fold
- Less than 2 Double-folds
- Less than 3 Double-folds
- More than 3 Double-folds

Paper Type and Condition (Mark all that Apply)
- White and Strong
- Yellowish or Tan
- Brown
- Glossy or Coated
- Calendered
- Ground Wood Paper
- Pest Damaged
- Volume Indicates it is Printed on Acid-Free Paper

(Alternate list if less detail is needed)
White and Strong
Yellowish or Tan
Brown

Mutilation and Patron Damage (Mark all that Apply)
- Pages Marked with
  - Pencil
  - Ink
  - Highlighter

Paper Clips
Dog-ears
Sticky Notes
Bookmarks left in Volume
Pages Torn or Cut Out
Animal Damage
Pages or Cover Stained with Food, Drink, or Water

Type of Binding
Publisher Binding
Publisher Paper Binding
Pamphlet
Commercial Case Binding
Commercial Laminated Cover Binding
Children's Book
Children's Book with Library Cover
In-House Binding

Condition of Binding (Mark all that Apply)
In Good Condition
Remain in Stacks
Needs Treatment
Must be Replaced or Withdrawn
Damaged Spine
Loose Joints
Damaged Inner Hinges
Damaged Paper Cover
Cover off Volume
Red-rot Leather
Title Worn Off
Title Label Missing
Call Number Worn Off
Call Number Missing
Volume Damaged From Lack of Support
Insect Damage

(Alternate list if less detail is needed)
In Good Condition
Remain in Stacks
Needs Treatment

Date of First Circulation
- Never Circulated
- Restricted Use Collection
- Same Year as Imprint Date
- 1 Year After Imprint Date
- 2 Years After Imprint Date
- 3 Years After Imprint Date
- 4-9 Years After Imprint Date
- 10 or More Years After Imprint Date

Last Circulation
- Previous Year
- Previous 5 Years
- Previous 10 Years
- Previous 25 Years
- None in Last 25 Years
- No Circulation History
- Restricted Use Collection

Number of Circulations in Last 10 Years
- 0
- 1-5
- 6-10
- 11-15
- 16-20
- 21-25
- 26 or More

Has the Book been sent out on Interlibrary Loan
- Yes
- No

Imprint Date
- 2000-
- 1995-1999
- 1990-1994
- 1985-1989
- 1980-1984
- 1970s
- 1960s
- 1950s
- 1930-1949

1910-1929
1890-1909
1870-1889
1850-1869
1830-1849
1800-1829
1750-1799
Pre 1750

Place of Printing
U.S.
Canada
Latin America & Caribbean
Great Britain
Northern Europe
Southern Europe
Eastern Europe (Former Soviet Bloc)
Former USSR
Northern Africa (Arabic States)
Central Africa
South Africa
Middle East
India
Central Asia
China and Asia (Not Japan)
Japan
Australia/New Zealand
Other (Add detail to Note field)

Language
English
French
Spanish
German
Italian
Russian
Asian
Arabic
Other (Add detail to Note field)

Previous Preservation Treatments (Mark all that Apply)

Damaged or Missing Pages Replaced
Been Repaired In-house
Conservation Treatment
In Acidic Box
In Acidic Pamphlet Binder
In Acidic Paper Envelope
Volume Tied Together with String
In Acid-free Box
In Acid-free Pamphlet Binder
In Acid-free Envelope
Dust Jacket Protector
Paperback Cover Stiffener
Been Reformatted

Shelving Condition (Mark all that Apply)
Shelved Correctly
Shelved too Tightly
Not Standing up Straight on Shelf
Shelved on Fore Edge
Shelved on Spine
Shelved in Wrong Location

How Many Copies of this Volume Does the Library Own
1
2
3
4
5
6

Treatment Decision for Volume (Mark all that Apply)
In Good Condition
Send to Stacks as is
Needs Commercial Binding
Needs In-house Repair
Needs Conservation Treatment
Send to Brittle Book Processing
Place in Enclosure
Must be Replaced or Withdrawn

(Alternate list if less detail is needed)
In Good Condition

Remain in Stacks
Needs Treatment

Barcode number

Call Number

Notes field

Instructions for each Collection Survey Question
In order for this condition survey to be valid it will be important for *all* surveyors to score the questions the same way. This can be difficult because so many of the questions are subjective. However, by carefully qualifying how each question should be answered, and by training the surveyors to answer questions the same way, consistency can be obtained. Through the years I have used these survey questions in over a dozen libraries around the world, and the data gathered is always constant if standard definitions and proper training are used. Therefore, it is important that every surveyor use the following criteria for each question when surveying the volumes in a collection. If a library decides to add additional questions to their survey, they need to carefully document what is meant by the question, and provide instruction to the surveyors about how to answer the question in a manner similar to the following documentation provided for the questions above.

Size of Volume

These sizes represent the shelving size location designations used by your library. Use whatever size descriptions are appropriate for your library. It is important to know the size of the volume being surveyed because size dictates how the volume is used and what kind of damage it will receive.

Regular

Volumes in regular stacks.

Folio

Larger volumes shelved by themselves in the "folio" stacks. Many libraries use a cut off of 30 cm to determine this cutoff. Some libraries use the Quarto size in addition to the Folio size.

Oversized

Very large materials, usually stored flat in "oversized" stacks.

Type of Volume

Designates what kind of a publication the item is. Again, use volume descriptions that will be most meaningful to your library. This will help in comparing damage to publication types, it will also help in estimating binding costs which differ for each binding type. This data can also be useful in determining information about how collections are used by patrons. For example, it will help determine how much monographs get used in a specific subject area. It can also let a library know how much their journal titles get used after they are bound.

Monograph

Single volume monograph.

Part of Multi-volume Set

Monograph that is part of a multi-volume set, series, and so forth.

Serial

Periodical literature, magazines, journals, and so forth.

Scores

Musical scores.

Leaf Attachment

It is important to determine how the leaves (or pages) of the text block are held together to know what kind of binding and preservation problems to expect with each type of leaf attachment method, and to determine future binding costs (e.g., recasing costs, adhesive binding costs, and extra handling charges for removing side sewing). This data can also be used to gather information about how your library collections are changing over time.

Sewn Through the Fold

Items with either machine or hand-sewn signatures. You can identify these items by looking for sewing thread in the gutter, or inner margin, of the middle of a signature, or gathering of pages. Just because a text block has signatures does not mean that it is sewn. *Do not* be fooled by burst-bindings which still have signatures, but are adhesive bound instead of sewn. Also, some volumes have signatures that are held together with staples (see below).

Oversewn

A long practiced leaf attachment method used to sew single leaves together—especially common in older commercial bindings. Individual leaves are grouped in small (about 1/8") gatherings, which are side sewn together. The oversewing machine operator continuously feeds these small gatherings into the machine which sews the gatherings together. Oversewing can be recognized by a very inflexible spine. Also, the inner gutter is tight and when the volume is forced open you will be able to see uneven sewing threads sewing the pages together.

Adhesive Bound

Text block is held together solely by adhesive. This is done by either gluing loose leaf pages together, or by gluing signatures together which is called a burst binding. Burst bindings are made by forcing hot melt adhesive into the folds of the signatures to hold the pages together. Adhesive binding methods are used by both publishers and commercial binders. It is now the most common form of leaf attachment used.

Stapled Through the Fold

Some books published in the late nineteenth and early twentieth centuries were made by stapling the signatures to crash or mull that was glued to the spine of the text block. Staples are also used in many pamphlets. It is important to know about staples because they can rust and damaged the paper.

Side Sewn or Stapled

Some volumes, particularly government documents and middle and far eastern publications, are made by stapling or sewing the pages or signatures of the volume together through the side of the text block. This is also a very common form of leaf attachment in children's books. This is a very damaging leaf attachment method for volumes with western papers because it makes the paper flex more than it wants to, causing it to eventually weaken and break—especially if the paper is older and brittle.

Spiral or Other Loose Sheet Binding

This category is for any material bound in three ring binders, spiral bindings, plastic comb bindings, or other type of nonpermanent binding.

Condition of Text Block

This will rate the condition of the text block (the book minus the cover). Paper condition and the binding, or cover, will be looked at later. This only considers the block of pages that make up the volume. It does not consider mutilation, brittleness, torn pages, and so forth. This question has some important theoretical implications about how well various leaf attachment methods hold up over time. It will also help a library determine what causes their volumes to fail—the cover or the text block. It provides information about the publisher bindings and the commercial bindings in your collection.

In Good Condition

The text block is perfectly sound. No breaks, it is not deformed, it is not loose or sagging out of shape. If this ranking is given, then the text block can have none of the following problems.

Remain in Stacks

The text block is not in good condition, but it is in good enough shape that it can remain in the stacks for further usage. In other words, the condition of the text block itself would not give cause to pull the item to be sent for preservation treatment. The primary determinant for this ranking is to answer the question: Will the text be damaged, pages lost, or repair costs increased by not pulling the volume now? If this ranking is given, then the text block can still hold up to use even though it may have one or more of the following problems.

Needs Treatment

The text block is damaged to the point that it cannot be used before it is repaired. This means that to use or circulate the volume would risk information being lost from the volume.

Must be Replaced or Withdrawn

The text block is damaged to the point that it cannot be used and it cannot be repaired. The type of library and its preservation facilities largely dictate what can be repaired and what cannot. For example, a library may find it better to withdraw an old damaged book rather than paying to have it treated by a conservator, which could cost hundreds of dollars. It may also decide to simply replace the volume with a new copy or a new edition. For research libraries, a new edition may not be adequate so they have to repair the damaged volume if possible. Answering these questions are key examples of how this survey must be conducted in the context of the library's collection development strategies.

Broken or Loose Sewing or Adhesive

The text block is still in one piece, but the threads are loose, the adhesive has broken down, or the signatures in the text block are loose. Another sign of this problem is if the volume opens to one spot where the spine is broken, but the text block as not yet split in two pieces. A volume in this condition may or may not be able to remain in the stacks.

Broken Text Block

The text block has actually broken into two or more sections, or clumps of pages are coming out of the volume. A volume in this condition will need preservation treatment.

Loose Pages

When single or multiple pages have detached from the text block. This will often occur as a result of a split text block, or because the adhesive in an adhesive binding fails, allowing pages to break free. Other times, paper can be so brittle that pages will break away at the inner margin. A volume in this condition will very likely need preservation treatment.

Damaged Pages (not mutilation)

This refers to any damage to pages—usually tears—that occurs as a result of normal use, or because of text block damage. However, this does not include mutilation or malicious patron damage such as writing on or marking pages, tearing pages out, and so forth. When in doubt, damage will be attributed to normal wear and tear rather than to mutilation. A volume in this condition will very likely need preservation treatment because if not fixed, a small tear can quickly become a big tear.

Missing Pages (not mutilation)

Pages that are missing as a result of non-patron damage. Mark this answer when pages fall out of a book after they become loose through damage rather than mutilation. If it is clear that the pages have been torn or cut out of the text block, then mark that it has missing pages ***and*** that the book was mutilated. When in doubt, damage will be attributed to normal wear and tear rather than to mutilation. A volume in this condition will need preservation treatment.

Pages Damaged or Curled from Lack of Support (Paperbacks)

This is a category specifically for paperback materials. This will allow you to measure how many paperback items have damaged pages because they do not have the support of a hard cover. This can be useful information in deciding whether or not to initiate or continue a prebinding program.

Multiple Pages Torn near the Gutter (Mostly Children's Books)

Torn pages near the gutter of the book is an extremely common problem for children's books. This happens for several reasons. First, children are small, and their books are often rather large, proportionally speaking. Therefore, kids tend to turn the pages by sliding a thumb under the bottom of the page near the gutter on the page they want to turn. They then try to flip the page over, and often the force of the thumb tears through the paper rather than turning the page. This happens because the force is quick and sharp, and because the paper is often weak in the gutter area because the text block is side sewn, causing the paper in the text block to suffer a lot of stress and damage in the gutter area where the paper is forced to bend excessively. These tears are difficult to mend effectively. If your library has a large children's book collection that is used by children (rather than college students studying elementary education), this is an important specific piece of information to record.

Mold Damage

Mold is a serious problem for many libraries and their collections—especially in warm, humid areas. Mold on a volume cover can often be cleaned off or rebound, leaving no trace of the mold, but if a text block gets mold on it, the signs usually remain forever. Some of these signs are a moldy, musty smell in the paper and mold-stained pages. Mold can leave stains of all different colors ranging from black, brown, yellow, or red. Usually a text block will only develop mold if it has gotten wet, so a good sign that you should look for mold damage in a volume being surveyed is if there are clear signs that the volume has been water damaged. A volume in this condition will very likely need preservation treatment.

Gutter Margin Width

As paper becomes more expensive, publishers are providing narrower margins in the volumes they produce, especially periodical and paperback materials. The inner margin, or gutter, needs to be wide enough to allow for future rebindings and for good readability and photocopying. Measure the gutter margin width with a ruler from the text printed closest to the gutter. Check through the volume to find the text printed closest to the inner margin. This question can produce interesting results, but may not produce information that is as directly relevant to the collection development or preservation planning efforts in your library.

Less than ½ Inch

More than ½ Inch, but Less than ¾ Inch

More than ¾ Inch, but Less than 1 Inch

1 Inch or More

Paper pH

The pH of paper greatly affects how long it will last. Acidic paper will, generally speaking, last between 60 and 150 years, while alkaline paper will last many centuries. The easiest way to test the paper's pH is to use a special marker type pen that contains chlorophenon red. These pens can be purchased from library preservation vendors like University Products, Light Impressions, or Gaylord. Make a half inch long mark on the paper near the gutter of a page near the center of the volume. When the mark dries read the color. Sometimes there are different kinds of paper in the same volume (e.g., often plates will be on a different kind of paper). Test the paper that makes up the majority of the volume. Test a page that is between other like pages (i.e., don't test a page that is next to a plate). Be aware that some off-white papers can make it difficult to read the color changes of the pH pen. It is important to know what percentage of your collections is printed on acidic paper so you can plan effectively for the long term.

Yellow or Clear (Acidic)

Yellow or clear means that paper is definitely acidic with a pH of 6.0 or lower.

Tan (Slightly Acidic)

Tan, or a faint, deep purple means the paper is slightly acidic with a pH range of 6.0-6.8.

Purple (Alkaline)

A rich purple or lavender means the paper is neutral or alkaline with a pH of 6.8 or higher.

Paper Fold Test

This is a test for paper brittleness. This test is made by folding a corner of a page over like you were dog earring it. Press the crease of the fold between your finger and thumb, and then fold the paper back the other direction and crease it again. This is one double-fold. Do this three times. It is important to perform this test on a part of the page that does not have print. It is also important to perform the test, when possible, at least three quarters of an inch into the page since most pages are more brittle along their edges than they are farther into the page. Non-ground wood paper (see "Ground Wood Papers" in the "Paper Condition" question) in books printed in 1970 or later generally does not need to be tested since paper that has been recently made will pass the three double-fold test.

Less than 1 Fold

Some papers are so brittle that they will not survive even a single fold before they break.

Less than 1 Double-Fold

The corner being tested breaks off after one double fold or less.

Less than 2 Double-Folds

Generally, in the preservation community, paper is considered brittle if it will not pass a two double-fold test.

Less than 3 Double-Folds

More than 3 Double-Folds

Paper Type and Condition

This looks at the physical characteristics of the paper in the volume. The detail of information gathered by this question may be more than many libraries need, and it takes some training and experience to be able to identify some of these kinds of papers or attributes. For example, it can be difficult for some people to tell the difference between calendered paper and coated paper, but for preservation purposes that distinction can be important. However, it is generally good to at least identify the color of the paper since this may have an impact on future reproduction of the page for either reformatting or for patrons copying or scanning the pages. In general, acidic paper turns yellow then brown as it deteriorates.

Environmental conditions can exacerbate the problem. High humidity can make the paper brown more quickly.

White and Strong

Paper is in very good condition like a new book.

Yellowish or Tan

As paper begins to deteriorate it turns first yellowish or tan and then brown.

Brown

This paper is generally further deteriorated than paper that is yellowish or tan. Obviously, it can be rather subjective to identify the difference between tan and brown. It is important to remember this when it comes to evaluating the data so that too much weight is not put on either of these ratings.

Glossy or Coated

Paper that has a shiny coating or is very smooth and glossy. Pages that have photos on them, art books, or many magazines have coated paper. Only mark this answer if the volume has glossy paper in the majority of the volume.

Calendered

This is paper that, at first, may appear to be coated, however, it receives its smooth texture not from a coating, but rather from being pressed by hot rollers when it was being made. It is, therefore, not shiny or glossy like a coated paper. Calendered paper was made primarily for lithography printed books of the late 1800s and early 1900s. Calendered paper is often seen in art books, and other highly illustrated books. The process of pressing the paper to make it hard for good lithography printing also makes the paper less flexible. As a result, calendered paper becomes brittle faster than other kinds of paper.

Ground Wood Paper

Ground wood paper is relatively inexpensive compared to high quality papers and is used often in newspapers and paperback books. It is also very widely used in printing done by underdeveloped countries. This paper is very acidic, is made from short, weak cellulose fibers, and has a high lignin content which causes the paper to severely brown when it ages (think of how yellow your newspaper gets after just one day in the sun). Many modern ground wood papers will test alkaline with the pH pen, and they will last much better now, but ground wood still lacks the physical strength of better-made papers.

Pest Damaged

Older materials, and materials from tropical climates, often have signs of pest damage. Often the signs are small holes in the paper that have been put there by "bookworms," which are beetle larvae that feed on the cellulose in paper. Cockroaches and silverfish will also feed on books and paper if they do not have other food sources. Cockroaches and silverfish will often nibble at the edges of pages much like a caterpillar eats a plant leaf. Often this will happen on the spine edge of a volume since the cockroaches and silverfish are attracted to the animal glue on the spine of some text blocks. It is very difficult to identify if a book is infested, so the best signs to look for are wormholes, bug eaten pages, dead larvae, "saw dust" from chewed up paper, or other obvious signs of infestation.

Volume Indicates it is Printed on Acid-Free Paper

Since the mid 1980s, most books (with the possible exception of some popular novel paperback volumes) printed in America and western Europe are printed on acid-free paper. The reason for this change is largely because of environmental regulations placed on paper manufactures. Alkaline paper is more environmentally friendly. However, in America, the library community has pushed hard for publishers to used acid-free paper. As a result, publishers will often indicate that the volume is printed on acid-free or alkaline paper either on the front or verso of the title page. If the book in hand was printed in 1980 or later, check to see if the publisher has indicated that or later, check to see if the publisher has indicated that the book is printed on acid-free paper. Some times they will indicate this with an infinity sign (∞).

Mutilation and Patron Damage

It is important to determine what percentage of the collection patrons have damaged through mutilation or neglect. It will also be nice to know what kind of damage occurs most often. These data are important because they report on damage that is completely preventable because the damage is caused by neglect, ignorance, or malicious intent. Knowing what kind of damage is occurring will help your library know what kind of educational programs to use to help prevent the damage.

Pages Marked with

Check quickly for any patron-made marks in the text and record what kind of marks were found. As a general rule, volumes are more heavily marked at the front of the book than in the back, so the front of a book is the best place to look. Also, if the text in the volume has been marked or underlined once, it is likely that it will have been marked several times, so if you find markings, check closely for more. Also, the more heavily circulated a volume is, the more likely it is to be mutilated in some way.

Pencil

The kind of marking in the text determines what can be done to remove it. Removing pencil marks, while labor intensive, is possible. Removing ink or highlighter is nearly impossible.

Ink

Ink and highlighter can make the text hard to read, and can cause problems for future reformatting efforts.

Highlighter

Paper Clips

Mark yes to this question if there are paper clips in the text block, or if you can tell, by damage to the pages, that there were paper clips used as book marks. Paper clips can damage pages and they can rust over time.

Dog-ears

When a corner of a page is folded over to mark a place. Mark yes to this question if you find folded over corners, or if you can tell that a corner was dog-eared from a crease left on a page. You can often identify if a page has been dog-eared by fanning out the text block a little and looking at the pages from the top of the book to see if any of the pages have been folded over. It seems obvious, but experience has shown that it is sometimes important to remind surveyors to not count the page that they just used as for the double fold test as being dog-eared.

Sticky Notes

Sticky notes can be damaging to paper because residual adhesive is left behind when the sticky note is removed which can stick pages together. More importantly, the adhesive used on sticky notes releases easily for a while, but in time, the adhesion becomes permanent which means the note cannot be removed without tearing the paper. This problem is exacerbated with weakened paper when the adhesive is stronger than the paper. Sticky notes can also lift print from the page.

Bookmarks left in Volume

People use all sorts of things for bookmarks and they often leave them in the volume when they are finished with it. Large objects like pencils, and very acid materials, like newspaper, can permanently damage the volume. Answer yes to this question if bookmarks, other than dog-ears, paper clips, and sticky notes, are found in the volume.

Pages Torn or Cut Out

Answer yes to this question if it is obvious that a page(s) was torn out of the volume, or if pictures or text were cut out on purpose. If there is doubt about how the page went missing do not count it as mutilation.

Animal Damage

For volumes that get chewed on by dogs, or for any other type of damage to a volume that is clearly the result of an animal.

Pages or Cover Stained with Food, Drink, or Water

Mark yes to this question if there is clear damage to a volume's pages or cover that is caused by food, drink, or water (e.g., rain). A clear sign of water damage is if the text block or cover is cockled or warped.

Type of Binding

The text block and paper of the volume have been examined, and now the condition of the cover will be recorded. These data are useful for determining how well various kinds of binding hold up to usage in your library. It could have a direct impact into what kind of shelf preparation efforts your library should make. For example, data from this question have led several academic libraries to conclude that they do not have to prebind all publisher paper bindings. However, since different libraries have different collection emphases and user patterns it is important for each library to gather this data for their institution.

Publisher Binding

Any volume that is in a hardbound cover produced by the publisher. Do not count commercial bindings here even if the book came prebound from your library book supply vendor.

Publisher Paper Binding

Any paperback volume that still has its paper cover.

Pamphlet

Any pamphlet, or single signature, saddle-stitched item—including musical scores. Record all pamphlets here, regardless of whether they have received an in-house pamphlet binding or not. It is important to know what percentage of your collection is made up of pamphlet-type materials. This will help in estimating shelf preparation costs for these materials.

Commercial Case Binding

Any volume that has been sent to the commercial bindery, or has been given a commercial binding by your book supply vendor as part of a shelf preparation agreement you have established with your jobber. A commercial binding can be identified by its heavy buckram cloth cover and its generally gold or white stamped label on the spine.

Commercial Laminated Cover Binding

A commercial binding that employs the original paper cover from a paperback volume by laminating it onto a paper backing and then using it to cover the new case binding created for the item. This is a popular binding method used in many public libraries, and is offered as a shelf preparation service by many book supply vendors.

Children's Book

The construction and use of children's books creates many preservation challenges for a library. It is important to know how well your children's books are holding up to use, and what percentage of your collection is made up of children's books. You may not need this answer if your library does not have a children's book collection or if small children do not use the children's collection. This answer applies to children's books, not young adult books that are bound like other novels.

It applies to volumes with publisher bindings, which often have paper-covered hard bindings with stiff spine inlays.

Children's Book with Library Cover

Because it has been recognized that children are hard on books, many vendors offer children's books with special covers specifically designed for libraries. These bindings may survive better than the publisher bindings, but do they protect the text block and pages any better? This is something that your library will need to determine.

In-House Binding

If your library has a conservation program that rebinds books, or if you do other types of in-house binding such as a shelf preparation operation, you will want to identify what percentage of the collection has been treated in-house, and how well these in-house volumes are holding up to use.

Condition of Binding

This section will record the condition of the binding, or cover of the volume. An earlier question recorded the condition of the text block. This question uses similar criteria to determine the soundness of the cover or binding. It is important to record data on the two major components of a bound volume in order to identify how well your collections are holding up and where failures are occurring. Depending on the purpose of the survey, it may be that your library does not need as much detail as provided below.

In Good Condition

Binding is in good condition, displaying none of the problems listed below. Cover appears almost new with no obvious signs that it is starting to fail in any way.

Remain in Stacks

This describes a book that is in good enough condition to remain in the stacks, but is not in perfect condition. It may have slight damage, but not enough to warrant sending it for preservation treatment or rebinding. The primary determinant in this case is to answer the question, will the text be damaged, pages lost, access diminished, or repair cost increased by not treating the volume now?

Needs Treatment

The binding is damaged to the point that it cannot be used before it is repaired. This means that to use or circulate the volume would risk information being lost from the volume.

Must be Replaced or Withdrawn

The binding is damaged to the point that it cannot be used and it cannot be repaired. The type of library, and its preservation facilities, largely dictates what can be repaired and what cannot. For example, a library may find it better to withdraw an old damaged book rather than paying to have it treated by a conservator, which could cost hundreds of dollars. It may also decide to simply replace the volume with a new copy or a new edition. However, for a research library, a new edition may not be adequate. If the paper in the volume is strong, then generally the volume can be rebound regardless of how damaged the binding is, but there may be times when this is not the case.

Damaged Spine

This is a situation where the spine of the cover is damaged enough to cause a lack of structural support, or to allow the labeling information to be lost, or to allow part, or all, of the spine of the cover to fall off. Damage usually occurs in two ways, 1) from the headcap being pulled on to remove a book from the shelf, and 2) from the spine inlay cutting through the covering material. If a spine is loose it will weaken the support of the cover and allow the inner hinges to become damaged very quickly. For some institutions with a good book repair facility, it can be cost effective to provide a spine repair on a volume before it becomes damaged enough to require rebinding.

Loose Joints

This describes the condition where the cover has become loose on the text block because the joint areas (the part of the cover that hinges) are loose. This happens because the weight of the text block pulls itself out of the cover. A volume falls into this category if it has loose joints, but has no damage to the materials that make up the inner hinges or outer joints.

Damaged Inner Hinges

The inner hinge is the joint area on the inside cover of the volume. The material in the inner hinge is usually paper which, being weaker than the cloth of the outer joint, breaks first. If the inner hinge of a volume is damaged it will quickly lead to other damage to the cover and text block.

Damaged Paper Cover

This question will help you gain information about how well paperback volumes hold up to library use and storage. Documenting this information can help shape collection development, shelf preparation, and preservation decisions. A damaged paper cover is one that no longer protects the volume, is delaminating, is breaking in the joints, coming off the text block, wearing away, or generally in bad condition. Include all kinds of paper covers such as paperbacks, and pamphlet bindings—anything with a damaged paper cover.

Cover off Volume

A cover of any kind that has broken away from the text block either in parts (such as a single board, or one side of a paper cover) or as a whole cover.

Red-rot Leather

The term "red-rot" is used to describe the dry, crumbly, weak condition of leather when it deteriorates. When you touch red-rotted leather your fingers often pick up a dark, dry powdery material. A good example of red-rotted leather can be seen on old tan, leather bound government documents or law books.

Title Worn Off

Mark this if the stamped titling information has been worn off of the spine of the volume.

Title Label Missing

Mark this if the titling label, from the original publisher binding or as a result of a repair, has fallen off of the spine of the volume. This is an important piece of information to record about the volumes in libraries that have a tradition of applying title labels to volumes that have been repaired or for other reasons. If your library has such a practice it is important to know how well those adhered labels are holding up to time and use.

Call Number Worn Off

Mark this if the stamped or written call number information has been worn off of the spine of the volume. This information may not be important for libraries that have a long tradition of applying call number labels.

Call Number Missing

Mark this if the call number label that was placed on the volume by library staff or a vendor has fallen off of the spine of the volume. This could be important information for determining how effective the call number labeling materials are that are being used in your library, or by your library vendor.

Volume Damaged From Lack of Support

This applies mostly to paperback volumes and pamphlets. It is important to know the percentage of materials in the collection that is damaged because they do not possess the necessary structural support to survive in the stacks.

Insect Damage

Cockroaches and silverfish will eat the starch-filled cloth covers of many books. Insect damage looks like something has scraped off the top layer of the cloth, leaving white spots. The damage can sometimes appear to be water spots on the cover, so look closely.

Date of First Circulation

Circulation histories can be very helpful in determining collection usage patterns and what kind of a treatment individual items need. If the item has not circulated very much, or if the item has not circulated recently, it will not need the same kind of preservation treatment that a heavily used item does. Circulation information can be taken from the date due slips in the back of the volume, or from the circulation database from the library's online system. If the barcode information is captured for each item surveyed, then a great deal of detailed circulation information can be obtained for each item to answer the following few questions. This is the preferred way to get the data if possible, but there will be some libraries that will need to rely on date due slips or other forms of circulation records if they do not have online records for all the materials in their collections, or if their online records do not go back very many years. This particular question is designed to see how soon an item circulates after it is acquired. The time ranges provided for this question can be set by the library depending on how detailed of information they want, and what their circulation records can provide. Circulation information is important because many items may circulate a great deal when they are new, but then circulate very little after a few years. Other items receive constant, heavy circulation all the time. It is important to know how various collections circulate and get used over time. This has implications for future collection development activities, shelf preparation activities, and preservation efforts.

Never Circulated

This answer, and several others, appear in most of the circulation questions. This data only needs to be gathered once, but it has been included in each question to remind the library to record this data at some point. The sophistication of your library online system and kind of information your library wants will determine what questions to ask and how best to get the data.

Restricted Use Collection

Same Year as Imprint Date

1 Year After Imprint Date

2 Years After Imprint Date

3 Years After Imprint Date

4-9 Years After Imprint Date

10 or More Years After Imprint Date

Last Circulation

This particular question is designed to see when the item was last used as apposed to first used. This is important because many items may circulate a great deal when they are new, but then circulate very little after a few years.

Previous Year

Previous 5 Years

Previous 10 Years

Previous 25 Years

None in the Last 25 Years

No Circulation History

Restricted Use Collection

Number of Circulations in Last 10 Years

The previous question measured when the item was last used, and this question measures how much the volume was used during the most recent ten years. This helps a library know if an item is being regularly used over a long period of time. An item that is slightly damaged, but does get used very much will not need the same level of treatment, and may not need treatment at all as opposed to an item that receives heavy use.

0

No circulations in the last ten years is not the same as never circulating at all. An item may have circulated a great deal when it was first added to the collection, but has not circulated at all in the last ten years.

1-5

6-10

11-15

16-20

21-25

26 or More

Some libraries—especially heavily used public libraries—may want to go higher than 25 circulations in the past ten years.

Has the Book Been Sent Out on Interlibrary Loan

This data can be very useful—especially for research libraries that want to know detailed information about what kind of materials are being sent out on interlibrary loan. This question can provide detailed information that some libraries would not be able to obtain any other way.

Yes

No

Imprint Date

The age of a volume helps determines the age of the materials in the collection. It can also give some clues about what kind of preservation treatment an item will need. Bookbinding and papermaking practices have drastically changed over the last several hundred years. For this reason it is important to see how binding structures and styles from various time periods are holding up.

2000-

Nearly all scholarly and hardbound books printed in America and Western Europe are printed on acid-free paper. However, fewer and fewer publishers are producing books with sewn text blocks.

1995-1999

1990-1994

1985-1989

1980-1984

Some publishers begin regularly using acid free paper.

1970s

1960s

1950s

1930-49

Depression and World War II era. Paper and bindings are generally of poor quality materials—especially items printed in Europe. Paperback volumes start to become popular.

1910-1929

A high percentage of the paper from this era will be brittle, especially poor quality papers.

1890-1909

Much of the paper from this era will be brittle.

1870-1889

Much of the paper from this era will be very brittle and brown.

1850-1869

Most of the paper from this era is acidic and will be very brittle, but this was an experimental era in papermaking history, and, as a result, there are many types of paper and different levels of quality. Some of the paper made in this era will still be handmade paper that will be strong and flexible. Also, paper from France and other parts of Europe will sometimes be in extremely good condition because they often added calcium filler to the paper to make it white. The calcium made the paper alkaline and has helped preserve it in beautiful condition.

1830-1849

Pre-acidic papermaking era. Paper is not generally as good as older papers, but often this paper will be white, flexible, and strong. This is when they started using automated case binding construction for producing covers. These bindings were not as strong as older bindings, so while the paper may be strong, the bindings will often be deteriorated.

1800-1829

As a general rule, the earlier the paper was made, the better its quality. This is because it is made with better quality materials and with fewer machines. This is true for bindings as well, but during this time they started to develop shortcut methods to produce bindings in a quicker, less expensive manner, and as a result you will find some bindings that are beautiful and strong, and others that are inexpensive looking and deteriorating.

1750-1799

Pre 1750

Place of Printing

Where a volume was printed tells a great deal about the preservation needs of that volume. Many under-developed countries print their books and periodicals on very poor quality paper and the bindings are often equally poor. The list of publishing locations can be as short or long as needed for your library, and can be changed to meet the specific needs of the library. For example, on surveys I have worked on in Eastern Europe, we had much more detail about eastern European countries and the former Soviet Union. We even broke down the selection to cities within a country.

U.S.

Canada

Latin America & Caribbean

Mexico, Central and South America, and the Caribbean islands. Argentina, Belize, Bolivia, Brazil, Chile, Colombia, Costa Rica, Cuba, Dominican Republic, Ecuador, El Salvador, French Guiana, Guatemala, Guyana, Haiti, Honduras, Nicaragua, Panama, Paraguay, Peru, Puerto Rico, Suriname, Uruguay, Venezuela, West Indies

Great Britain

This gets its own listing because in many libraries a large percentage of the English language books are printed in England—especially older materials.

Northern Europe

Austria, Belgium, Denmark, Finland, France, Germany (West Germany), Greenland, Iceland, Ireland, Luxembourg, Netherlands, Norway, Sweden, Switzerland

Southern Europe

Greece, Italy, Portugal, Spain

Eastern Europe (Former Soviet Bloc)

Former communist or Soviet Bloc countries including Albania, Bulgaria, Czechoslovakia (Czech Republic, Slovakia), East Germany (GDR), Hungary, Poland, Romania, Yugoslavia (Bosnia, Croatia, Macedonia, Serbia, Montenegro, Slovenia)

Former USSR

Russian Empire until 1917 when it became the Soviet Union until 1991. Armenia, Azerbaijan, Belarus (White Russia), Estonia, Georgia, Kazakhstan, Kyrgyzstan, Latvia, Lithuania, Moldova, Russia, Tajikistan, Ukraine, Uzbekistan

Northern Africa (Arabic States)

Algeria, Egypt, Libya, Morocco, Tunisia

Central Africa

Angola, Benin, Botswana, Burkina, Burundi, Cameroon, Central African Republic, Chad, Congo (Brazzaville), Djibouti, Equatorial Guinea, Eritrea, Ethiopia, Gabon, Gambia, Ghana, Guinea, Ivory Coast, Kenya, Lesotho, Liberia, Madagascar,

Malawi (Nyasaland), Mali, Mauritania, Mozambique, Namibia (South West Africa), Niger, Nigeria, Rwanda, Senegal, Sierra Leone, Somalia, Sudan, Swaziland, Tanzania, Togo, Uganda, Zaire (Belgium Congo), Zambia (North Rhodesia), Zimbabwe (South Rhodesia)

South Africa

Middle East

Iran, Iraq, Israel, Jordan, Kuwait, Lebanon, Oman, Saudi Arabia, South Yemen, Syria, Turkey, United Arab Emirates, Yemen. Depending on the collection, your library may or may not want to include Israel in this list. If your library has a large Judaic collection you may want to have Israel as its own selection. Also, in terms of book manufacturing traditions and structure, most books produced in Israel are more similar to European books in construction than they are to Arabic country publications.

India

Central Asia

Afghanistan, Bangladesh, Bhutan, Burma, Nepal, Pakistan, Sri Lanka

China and Asia (Not Japan)

Cambodia, China, Hong Kong, Indonesia, Laos, Malaysia, Mongolia, North Korea, South Korea, Papua New Guinea, Philippines, Thailand, Vietnam

Japan

Australia/New Zealand

Other (Add detail to Note field)

Language

The country of publication gives a strong indication of what language the publication is in, but it is not a perfect match. For larger research libraries this information can be very important. It is also important for area studies collections so they can see what percentage of their collections are made up of different kinds of languages. Again, the collection being surveyed will determine what languages to include in your list.

English

French

Spanish

German

Italian

Russian

Asian

Arabic

Other (Add detail to Note field)

Previous Preservation Treatments

It is important to know what percentage of the collection has received previous preservation treatment. This lets your library know what kind of treatments have been used in the past, how well they are holding up to patron use, and will help identify weaknesses in treatments currently being used. The list provided below can be changed to meet the needs of your library. For example, I worked in a library where tens of thousands of items were placed in poorly constructed temporary bindings. If I were surveying the collections in that library I would want to include that as one of the answers for this question.

Damaged or Missing Pages Replaced

Mark this if photocopied replacement pages have been tipped into a volume.

Been Repaired In-house

Mark to indicate that a volume has received some kind of an in-house repair. This can include paper mends, spine repairs, hinge repair with cloth tape, or minor conservation treatment if your library has those capabilities. Depending on how detailed of information you want, you can break the book repair and conservation answers out into several answers to match the treatments performed in your library. For example, instead of one broad answer of "been repaired in-house" and "conservation treatment" you could have answers such as spine repair, case binding, reback, paper mend, and so forth.

Conservation Treatment

This answer only needs to be in the survey if your library has a conservation program that is producing professional level conservation treatments for the collections you are surveying. If this is true, it is important to have this information in order to get an idea of what percentage of the collection has been treated, and how those treated items are holding up.

In Acidic Box

If the volume has been placed in some kind of a protective box that is made of acidic materials. Acidic materials can be identified by a dark brown board, or from acid burns on the paper enclosed in the box, or by age. Anything older than the early 1980s will be acidic. When in doubt, test box material with pH pen.

In Acidic Pamphlet Binder

Pamphlet binders that match the same criteria as given above. When in doubt, test binder material with pH pen.

In Acidic Paper Envelope

Many libraries have used envelopes to hold damaged books together so they can return to the shelf. If your library has not used this practice you can delete this answer from the list. When in doubt, test the envelope with pH pen.

Volume Tied Together with String

Many libraries use string, red cotton tape, or rubber bands (a very bad practice) to keep the parts of damaged books together so they can return to the stacks.

In Acid-free Box

Boxes that were made since the mid-1980s should be acid free. Many high quality boxes will have the pH stamped onto the box. When in doubt, test box material with pH pen.

In Acid-free Pamphlet Binder

Pamphlet binders that were made since the mid-1980s should be acid free. When in doubt, test binder material with pH pen. Be aware that some binder covers are too dark to allow for good test results, and some modern board has an acrylic coating that will prevent you from getting a good pH test.

In Acid-free Envelope

Usually, these are thick, white or tan paper envelopes that will not have adhesive on the envelope flap. Tyvek envelopes are also acid-free. The pH pens cannot test the pH of Tyvek.

Dust Jacket Protector

Dust jacket protectors provide several important functions. They protect the dust jackets for use by patrons in selecting a title. As a general rule, dust jacket flaps are the only source of information about the author of the book, and are the only place where a summary of the book can be found. Dust jacket protectors guard the covers of the books from abrasion and damage. Books that do not have dust jacket protectors will regularly develop cover damage on the spine because patrons and library staff often remove volumes from the shelf by pulling on the head cap or hooking their fingers in the hollow of the spine. A plastic dust jacket protector supports the spine of the cover and prevents damage. The plastic also protects the book from damage by water or other spilled liquids.

Paperback Cover Stiffeners

Several different products exist for stiffening and protecting paperback covers. Some of these products work very well, while others are detrimental to the volume. If the plastic material used to stiffen the cover is too stiff it will literally tear the paperback cover off the text block. Unlike dust jacket protectors, this process is not reversible. Furthermore, it involves applying a very sticky adhesive to the volume. The adhesive can ooze, or the stiffening material can separate from the paperback cover—especially if a treated volume gets too near a heat source like direct sunlight or a heating vent. This can result in a sticky mess that attracts dirt and leaves a sticky residue on everything the book touches.

Been Reformatted

Since the survey only looks at volumes in the stacks, reformatting refers to preservation photocopying. If your library does not do preservation photocopying, then you may not need reformatting as a potential answer to this question.

Shelving Condition

This question can be very useful for gaining information about how volumes are stored on the shelves. Many libraries work very hard to shelf read their collections and to keep their stacks in good order, but if a library has open stacks, then it is a constant struggle to keep the shelves in good shape. Other libraries find that with limited budgets or staffing shortages they have to reduce the amount of shelf reading they can do. Therefore, it is even more important for such a library to know the shelving condition. Depending on how you conduct the survey you may need to devise a small flag to record this data for each book. This will be discussed in the next chapter about conducting the survey.

Shelved Correctly

Volume has been shelved correctly. That is, it is in the right location, it is not too tight on the shelf, is not too tall for the shelf, and so forth.

Shelved too Tightly

Volume is on a shelf that is packed too tightly making it hard to remove it from and return it to the shelf. A volume is shelved too tightly when you cannot remove it from the shelf without dragging adjoining books off of the shelf with it. If the surveyed volume is too tight, then the entire shelf is packed too tight. This means that every volume in your survey that is shelved too tightly needs to be multiplied by 10 to 30 times depending on the average number of volumes held on your library's shelves, which shows just how big of a problem tightly packed shelving can be.

Not Shelved Straight

Shelf is too loosely packed, allowing the volumes to lean or fall over. It is very damaging to a volume to have it leaning on a shelf. It weakens the binding by stressing the joints. If a book is not shelved straight it may mean that one or more of the volumes shelved next to it have been removed by patrons, allowing the surveyed book to lean. Or, it is often the case that if one book is not shelved straight that many of the books on the shelf might be in the same condition.

Shelved on Fore Edge

Volume is too tall for the shelf and is, therefore, stored on its fore edge with the spine facing up. This is extremely damaging to the book and will cause gravity to pull the text block out of the binding. Libraries like to store tall books on their fore edges because it makes it easier for the patrons to read the spines, but the cost for this ease of access is damaged books.

Shelved on Spine

Volume is too tall for the shelf and is, therefore, stored on its spine—the fore edge facing up. It is best to have the shelving arranged to accommodate the height of the volumes stored on the shelf, but if volumes cannot be stored standing up properly, this is the proper way to store those that are too tall for the shelf.

Shelved in Wrong Location

Volume is not shelved in the correct location in the stacks. This could represent a problem with the reshelving staff, or it could simply indicate that patrons are removing volumes from the shelf and returning them to the wrong location as they browse the collection. This problem is usually higher in heavily used parts of the collections and in the literature sections where the volumes have long call numbers, making it easier to misshelve them.

How Many Copies of this Volumes Does the Library Own

This information could be very useful for a library that is considering a weeding project. Each library will have to decide how it wants to answer this question. Research libraries will probably want to only count exactly matching editions of the volume. Public libraries may be happy to count copies of the title regardless of edition information. How you get this information will depend on your library. Some libraries mark this information on the call number label. If so, you will have to get this information at the time that you pull the book. For this reason, it is put here in the list of questions after the shelving condition question. However, if your library decides to get this information from the library system, then you can slot it by the barcode questions or the circulation questions.

1

You can answer this question in two ways. You can either have a pull-down menu with numbers, or you can have an open text field in which to type a number. You can define the field to only take numbers. This is probably the better option because otherwise you may not have enough numbers in the pull-down list to cover all the copies of a volume that your library owns.

2

3

4

5

6

Treatment Decision for Volume

The information recorded by this question provides an overall preservation assessment of the volume in hand. It provides the most general preservation information of any question. It is, in a sense, a summation of several of the other questions. For example, if you recorded that the text block needed treatment, then you will need to record here what kind of treatment is needed be it commercial binding, conservation treatment, or reformatting. For this question, the answers you list need to be supported by the preservation services available to your library. For example, if your library does not have a conservation program, and will not develop one, then you may not want to have an answer about the volume needing conservation treatment. However, if your library has a conservation program, or if there is a possibility of developing one, then it will be good to know what percentage of the collections need conservation treatment. After all the other questions have been answered for the volume in hand, the book must finally be examined the same way your library's preservation staff would examine it to determine what treatment it needs. If your library does not have such a department, then use the same criteria that staff use who make preservation treatment decisions. One determination that needs to be made prior to the review process, and must be consistent throughout the survey, is how you will decide when to say an item needs preservation treatment, and when to say it can be sent to the stacks as is. You can either base the decision on current treatment decision-making practices used in your library, or you can base the decision on more idealistic criteria by deciding on what treatment the item really needs even if your library normally would not have the funds to provide the ideal treatment for most damaged items. Using current library practices gives the library a good idea of what it would cost to treat the damaged materials in the collections at the current level of support. This is useful information. Using more ideal standards could be useful for future planning to help grow preservation efforts.

In Good Condition

Volume is in good condition. Has nothing wrong with the paper, binding, and so forth.

Send to Stacks as Is

Volumes in this category do have some preservation concerns. They may have slightly loose covers, or slightly brittle paper, but nothing that will prevent it from going to the stacks or from being used one or more times before needing treatment.

Needs Commercial Binding

The decision to commercially bind a volume is largely based on the condition of the paper (paper must be able to pass a three double-fold test). Must have sufficient inner margin (3/8 inch or more). And have damage beyond what can be treated economically with an in-house repair (see below), and less damage than would warrant conservation treatment (see below). Items should be sent for either binding or repair if their structure is such that the item will not withstand use, or storage in the stacks (e.g., pamphlets, some spiral bound materials, side-stapled or sewn materials that do not open well, thin paperback materials, very large paperback materials).

Needs In-house Repair

A volume needs in-house repair if its pages need to be slit open, if it is a pamphlet that has not been placed in a pamphlet binder, if it has torn or missing pages, if it needs an enclosure, or if the spine of the cover is damaged, but the inner hinges are still sound. The level of book repair services available in your library will determine whether it is more cost effective to repair things in-house or to send them off to a commercial binder. This data will be useful in your library's treatment decision-making process. It will help you see what percentage needs repair, and what percentage needs binding, and it will help you decide whether or not you have these decisions properly balanced in terms of cost, time, and staff resources.

Needs Conservation Treatment

A volume needs conservation treatment if it is damaged and too old or fragile to be sent to the commercial binder, or if it has paper that will not pass a three double-fold test, but will pass a two double-fold test, or if it is a special item needing special care (e.g., special construction, no inner margin, needs lots of paper repair, text block needs to be resewn), or if it is a special collections item.

Send to Brittle Book Processing

If the paper in the volume will not pass a two double-fold test, and the text block is broken, it will need to be reformatted to be used in the future. Also, if the paper in the volume will not pass a two double-fold test and the text block is sound, but the volume has been used regularly in the past few years it should also be reformatted. Any volume that has paper that will not pass at *least* a single double-fold test will probably need to be reformatted or withdrawn.

Place in Enclosure

If the paper in the volume will not pass a two double-fold test, and the text block is fairly sound, and if the volume does not circulate regularly, it can be placed in a custom enclosure for its protection until it is determined, through future circulations, that the volume should be further treated. This is a particularly good option for journal volumes or multi-volume sets that would be prohibitively expensive to reformat.

Replace or Withdraw

If the book is too damaged to be repaired or not worth the cost of repair or reformatting it will have to be withdrawn from the collection. The library can then decide if it wants to purchase a replacement copy. These decisions are driven by collection development policy as much as they are preservation decisions. A large research library will go to great efforts to maintain a copy of a volume—even if it is expensive to do so. A public library will be more likely to replace the volume with a newer edition. Your library might want replace and withdraw to be two separate answers to distinguish the difference for collection development costs. This may be especially true if the bibliographer in charge of the collection being assessed is conducting the survey.

Barcode Number

Recording the barcode number is a fast, easy way to track every book surveyed. It also allows quick access to the online record for the volumes surveys so that additional bibliographic and circulation information can be pulled into the survey for analysis. The other nice aspect of recording the barcode information is it allows the library to repeat the survey five to ten years later with the exact same volumes so the library can compare how the volumes have aged, what additional damage has occurred, whether some of the volumes have been lost or withdrawn, and so forth. Access and other database programs enable you to use a barcode scanner to wand the barcode number directly into the database. These programs will then enable you to print out the barcode number in the form of a barcode to make reading the numbers easier in the future.

Call Number (Optional)

Place the call number of the item in the space provided if there is something about the volume that will necessitate it being retrieved at a future date. If you record the barcode information you probably do not need to also record the call number. You can simply make a note in the note field described below. If there are questions regarding the surveying of a volume, or if you want to be able to retrieve a surveyed book later, you can record the call number to help you find it again.

Notes Field (Optional)

This unrestricted text field is for use by surveyors to note things about the volume they are surveying if the need arises.

Chapter 3

Conducting the Survey

Developing an appropriate survey instrument that gathers the information needed by your library is a very important part of any collection condition survey. Equally important is the statistical part of the project. If the survey is not conducted properly, the resulting information will not accurately reflect conditions in the collection, and you will end up making strategic decisions based on faulty information.

Statistics are challenging. Statistical literature is full of familiar terms such as sample, distribution, mean, mode, and median that are used in unfamiliar ways. There are many well-written articles and books however, that can help you in structuring your survey and evaluating your data in a statistically valid way.[1] There are even good texts on library sampling.[2] However, it is generally a good idea to get the advice of an expert for the statistical part of the survey. Put together your survey, decide what information you want, then find an expert who is willing to review your work and give you advice. Every college and university will have statistical experts who can assist you. Also, most social scientists must have extensive training in statistics to do their work; it should be relatively easy to find an expert who will volunteer his or her services or offer assistance at a reduced price.

There are some general rules that can help you in the initial preparation of your survey. First, the size of your survey sample will be determined by how detailed the information is you are gathering. For example, political polls are valid to within a few percentage points with sample sizes in the hundreds. It does not matter that there are over 260 million people in the country. A sample of a few hundred people will give the same results as a survey of thousands. That is the beauty of statistics. This is possible because of the relatively homogeneous nature of the population and the relatively few choices in how the questions can be answered. By contrast, a survey to find out how people around the world feel about a United Nations action would require a larger sample because the world population is not as homogenous as the population of an individual nation. Furthermore, if the survey were intended to identify how various nations felt about a U.N. policy, it would take a much larger sample so that enough data could be gathered about each individual nation to report the findings at that detailed level.

The survey questions from the previous chapter gather a lot of detailed information about every volume sampled. This detailed information will allow a library to draw some very strong and meaningful conclusions about the make-up of their collections if a sufficient number of items are sampled.

In most small academic and public libraries, a sample of about 250 to 500 books will be sufficient to provide statistically valid information depending on how detailed the questions are in the survey used. A sample of 500 volumes will even allow for some cross analysis such as knowing if paperback novels circulate more than hardbound novels. However, if a library has several locations, and only 500 volumes are sampled across all the locations, then it may not be possible to compare findings from the various locations. When we conducted a condition survey at the University of Kansas Libraries, we sampled at least 350 volumes from each of the seven libraries we surveyed to enable us to compare results across locations. We also used a stratified sampling technique so that we sampled a much larger number of volumes from the larger libraries to ensure that the data from the larger libraries held the same weight and ability to predict the condition of those collections as did the data from the smaller libraries. As a result we surveyed over 3,675 items to have a large enough sample to provide us with the predictive power we wanted. By contrast, I have been involved with surveys of very large libraries where we surveyed less than 500 volumes from all their combined collections. This provided sufficient data to report on the nature of the library's collections as a whole, but not enough to report on individual collections. The detail you want will determine how many volumes you need to survey. Again, this is why it is important to know what level of information you want and how much cross analysis between questions in the survey and between the collections in your library. Once you have your questions written, your answers created, and have decided on the level of detail that you need, then you can determine your sample size. There are rules for selecting sample sizes, but this is where you really need the help of a statistician. Having the survey prepared and knowing what detailed information you want will help you explain your project to the statistician, and will aid the statistician in helping you pick a proper sample size. However, in general, a sample of 500 volumes will provide most libraries with sufficient statistically valid data to determine the makeup and condition of their collections.

After it is determined how many items must be sampled to provide you with the information you need, you can decide how best to collect your sample. The key point is to ensure the sample is randomly selected. This means that each item in the collection has an equal chance of being selected for the survey.

There are various strategies that can be employed when selecting a sample. One way is to pull a random sample of titles from your library's online catalog using a database program. It is relatively easy to design a query that will select a random sample from the online catalog and pull the title, author, call number, and other information from the catalog into the database. Then simply survey the selected titles. One advantage to this method is it gives you bibliographical information on every title that is surveyed. Such data can be useful in conducting longitudinal studies (the process of collecting data on the same population over a long period of time) on your collection. By surveying the same titles five or ten years later you can compare the two results and learn things about your collection you could learn no other way. However, it is important to remember that you can only use this method if *all* of your library's holdings are in the online catalog, otherwise your sample will be skewed.

Another easy way of selecting a random sample of your collections is to use a rule or set procedure for selecting items. A good way of selecting a sample is to determine how many items you need, how many shelves there are in the library, and then divide the number of items needed into the number of shelves. This can be done by using the following formula:

$$n = \textit{Number of shelves in a location} \div \textit{number of sample items needed}$$

For example, say you needed 300 items for your sample and your library has 12,382 shelves. To get a valid random sample you will need to select a book from every forty-first shelf, so $n = 41$. This will result in surveying more than 300 volumes, but not too many more. Remember, when selecting a survey sample the important number is always given as a minimum, not as an exact number. If you need to survey 300 books, 325 will not make your survey more valid, but fewer than 300 will statistically weaken your results. The reason we count shelves rather than ranges is because the stacks may have differing numbers of shelves per range. Counting shelves helps guarantee every volume in the collection has an equal chance of being selected.

Next, devise a rule for selecting an item from the appropriate shelf. Some people measure from the left or right side and pick the item that is twelve or fifteen inches from the edge. This works, but I prefer to count books. Generally, I will select the fourth or fifth book from the left edge of the appropriate shelf. I use a number like four or five because too high of a number like ten or twelve will require a great deal more time for the surveyors because they will have to count two or three times as many books on each shelf before identifying the needed volume. I also always count from the left hand side of the shelf because items are shelved from left to right, leaving the empty space on a shelf on the right.

It is important that you have well-defined rules for how you sample your collections, and that you always follow those rules. You need to clearly identify how you will select the surveyed items and stick to the rules throughout the survey. Your rules must address what you will do if the selected shelf is empty or if the selected shelf has less than the required number of books. For example, my survey rules often instruct the sampler to count empty shelves, but that if a shelf selected by the rule is empty to move on to the next available shelf with books. If my sampling technique requires me to take the fourth book and a shelf has only three books on it, my rule states to move to the next shelf and take the first available item. However, a rule to take the first available book to the left if there are not four books available would be just as valid. The important thing is to have a rule and stick to it. This is what makes it a valid sample.

How you physically decide to conduct the survey is another important decision. Under some conditions it is easier to pull the survey items from the stacks, place them on a book truck, and take them to an office or work space to conduct the survey. In this case, if you are gathering information about the shelving condition you will have to record this data at the time you pull the volume. An advantage to doing it this way is that you can have the volumes on a truck in your office and work on them as you have time. Also, if you are recording barcode information it is often easier to work at a desktop computer where you can employ a barcode reader to gather than information.

If the survey is being conducted in a building that is some distance from your office, or from a useable desktop machine, it may be easier to conduct the survey in the stacks using a laptop computer or a handheld computer. Also, if the library has a wireless network, it may make it easier to survey the items right in the stacks. Use technology—and its limitations—to help you determine how you will conduct the survey.

Organize the survey form in a way that will make surveying easy and efficient. Survey every item the same way, answering each question in turn so that you develop a pattern and rhythm. Organize the flow of the questions so they proceed from the outside of the book to the inside. First you identify the type of book, then the type of binding, and then you get the data you can gather off of the title page such as place and date of publication. Next you open the book to learn about the text block and the paper. Finally, you end with the circulation information—especially if this information is being taken from a date due slip in the back of the volume. Having a well-documented survey method will be important for training the survey team, for ensuring that the same procedures are used by all of the surveyors, and for future documentation in case the study is ever repeated—especially in the case of a longitudinal study. Having a well-laid-out survey form that flows in a logical way will also help make the surveying easier, and will reduce the likelihood of a question being missed in the survey—requiring you to go back and complete it when the survey form will not clear (if you are using a database form to conduct the survey). It will also result in survey questions being answered in a more standard, methodical way that will produce better overall results.

One final point in conducting the survey: use technology to help you. When properly used, technology can be extremely beneficial in conducting any library condition survey. There are many wonderful technological advances taking place in the computer industry. Barcode scanning and handheld computer technologies are developing rapidly. By the time this book is published, there will be new technology that your library can use to help you in surveying your collections. However, use the technology appropriately. A general rule for using technology in your survey is that simpler is better. The more complicated the procedures or system, or the more technology that is involved, the more potential there is for something to go wrong.

Creating a Survey Tool

The most effective way of recording and analyzing the data from the survey is to use a database program. Using database software can be a bit confusing at first. Setting up tables and survey forms involves thinking about things in new ways. It is not unlike learning mathematics or a new language. At first it is hard and confusing, but then it starts to make sense and you begin to understand how database programs work and how they organize and analyze data. Often people who are very comfortable working with computers and use them regularly will still be frustrated when first using database software. Some computer experience does translate over, but much does not. Being aware of these frustrations before you begin will help you stick to the project through the tough beginning stages. The first survey project is always the hardest because there are so many new things to learn and bugs to work out of the system, but you will find that with each survey project setting up the database, creating the survey forms, and analyzing the data gets easier and faster. In addition, the experience gained by library staff in using database software to analyze the library's collections will carry over into other areas of the library. Staff will discover new ways of gathering information and assessing services that will result in improved services to the community.

I have the most experience with Microsoft Access, but other database programs will also work. Create a table with a column for each question in your survey. If the question can be answered using a pull-down menu, then one column will be enough for the question. However, if there is a potential for more than one answer for the question such as questions that instruct the surveyor to mark all the answers that apply, then each potential answer becomes a yes or no question, and each answer will need its own column. Also include a column that will automatically number the records so you can keep track of how many items you have surveyed.

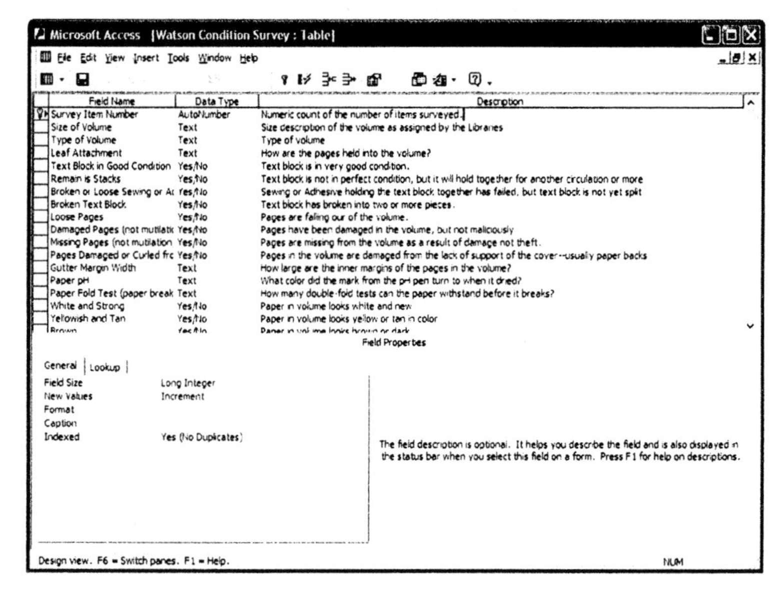

Figure 3.1: Setting Up an Access Table

Figure 3.1 shows how the table is created. Text put into the description field on the right describes the question or the specific answer to the question. This information appears in the status bar when you select the question or answer in the survey form later.

After the table is established, use it to create the survey form. If you are using a handheld computer to conduct your survey, you will probably still want to create a table in Access or some other sophisticated database program and then transfer the table to a handheld computer. Having said that, be aware that the software programs for handheld computers, and the handhelds themselves, are developing rapidly and becoming extremely sophisticated, and that by the time you read this book there may be seamless database programs that go easily back and forth between the desktop computer and the handheld computer. If you use a tablet PC to conduct the survey, you can run the same software packages and attachments (like a barcode scanner) as a computer while, at the same time, having the ease of use as a handheld computer.

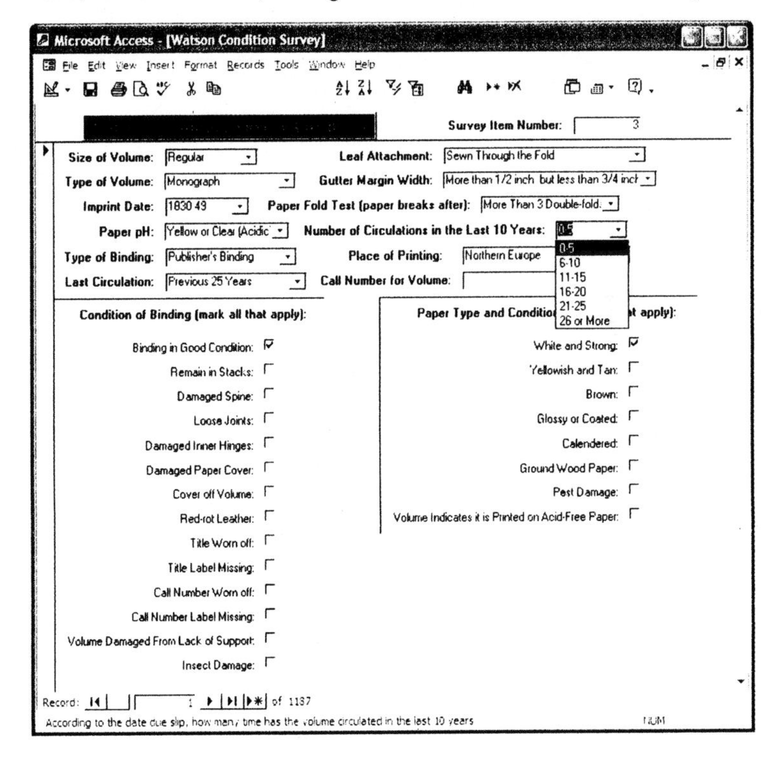

Figure 3.2: Example of Access Survey Form

The form is the main tool used to collect the data, so it is worth spending some extra time to make sure it is correct and easy to use. It is important to design the form to flow in a logical way both to go through the questions so that they flow through the book from outside to inside, but also so that the questions appear on the screen in a logical, condensed way. The more times you have to scroll down the page, the longer the survey will take.

Figure 3.2 shows how the survey form works. Questions with pull-down menus are kept together to reduce the amount of screen space taken up. For these questions, you can set up the table to require an answer. This provides a safeguard against forgetting to answer a question. If one of these questions is not answered the form will not clear. The same can be set up for free text fields like the barcode field.

The questions with more than one potential answer have the answers listed with a check box next to each answer. The default for each of these yes/no questions is no so that later you only have to query the yes answers. You can place page breaks in the form anywhere you want, enabling you to hit the page down key to move to the next set of questions. You can also assign the tab order to go through the questions in any order you want so that you can tab from question to question if you want. However, my experience has shown that it is easiest to simply work with the mouse with a wheel button on it to help you easy scroll down the form. If you use a laptop computer, it will probably be worth it to use an attached mouse rather than using the built-in touch pad or other form of built-in mouse.

A word of caution about Access and some other database systems: these programs are extremely powerful and flexible. The negative impact of such flexibility is that the program will allow you to easily change or erase data you have collected throughout the survey process. To guard against such accidents, make sure that you back your data up regularly. Also, be careful when using the data entry form to make sure that you are not recording new data over old survey records. With many of these programs, each time you open the form it automatically takes you to the first record in the table. The natural inclination is to open the form and start recording data, but make certain that you start with a blank record before you start to enter new data.

Using Handheld Computers

Using a handheld computer can save a great deal of time and effort in conducting the survey under the right conditions. Using a handheld allows you to survey items in the stacks rather than gathering them to a central location to enter the data onto a computer. Using handhelds is particularly advantageous for remote locations, stacks with poor lighting, cramped stacks, or locations where there are no work areas available. Using a handheld allows the surveyed volumes to be returned to the shelf in a completely non-invasive manner so that shelving staff do not have to reshelve hundreds of volumes.

Data entry forms similar to the ones shown in figures 3.3 and 3.4 can be created for handheld computers as well. These forms can have pull-down menus and yes/no check boxes. You can even have free text note fields.

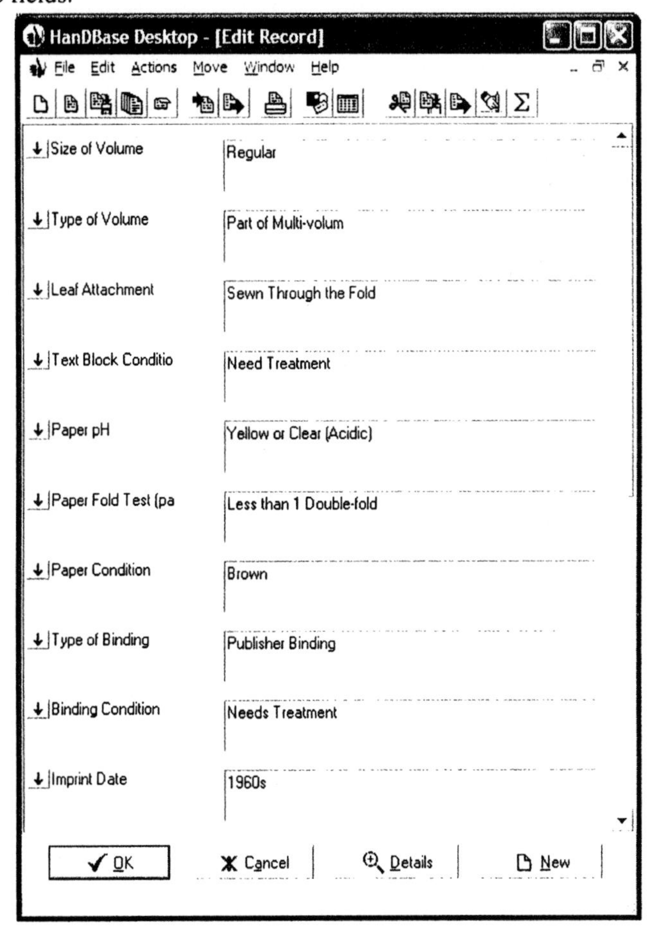

Figure 3.3: Example of PDA Survey Form

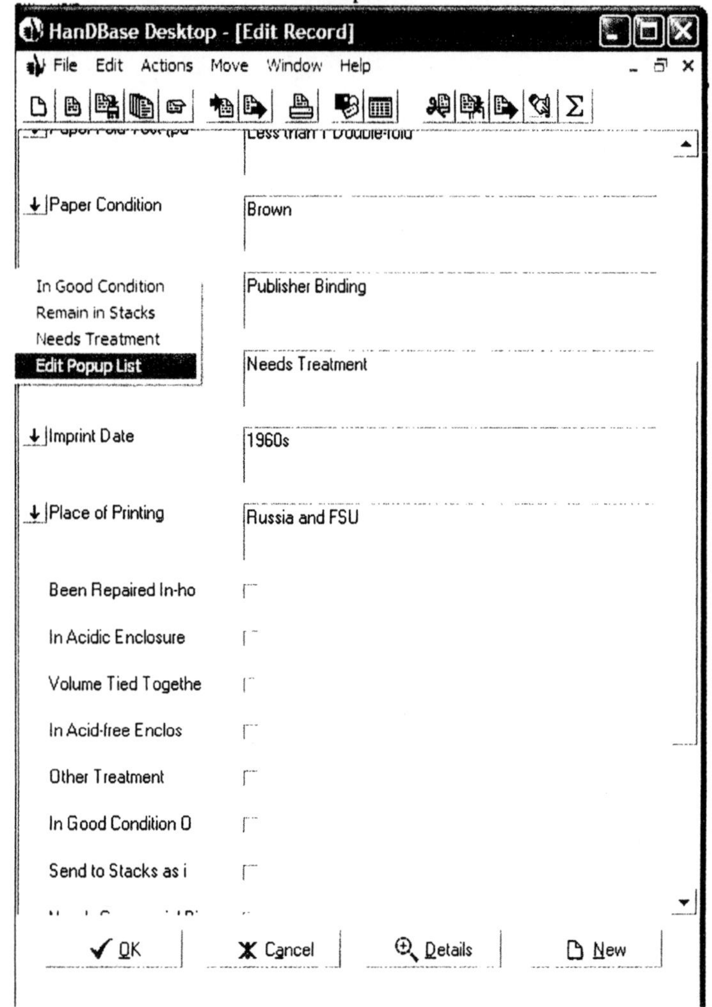

Figure 3.4: Example of PDA Survey Form

However, because the screens on handhelds are small, a long survey may become cumbersome to work with on a handheld. Many handheld database software companies also provide very nice form creation software that make forms easier to work with on the handheld. These extras are not needed, but they are nice—especially for handhelds with color screens. The other draw back at present is that current handheld database programs do not allow for you to easily scan barcodes into a database field. There are several handheld models that have built-in or added features that allow you to scan barcodes, but they generally need an interface program to work the scanner that records the barcode data in a list or table. Again, technology changes rapidly, so it will be worth checking to see if things have changed, but at present, if you plan to scan the barcode into the database, it may not be worth it to use a handheld computer for the survey. However, if you select a random sample of books from the online system, and then pull those books from the stacks, using a handheld computer would be ideal.

Also, you might want to consider using a tablet PC to conduct the survey. A tablet PC has the benefit of having the power of a PC and can run the same software packages and attachments (like a barcode scanner) as a computer while, at the same time, having the ease of use as a handheld computer. They can use a stylus to enter data directly into the form like you can with a handheld, but it has more power. Currently the tablet PCs have a battery life of a few hours, but the next generation of tablet PCs will be much lighter and have batteries that last up to ten hours of use.

Time Requirements

Depending on the number of questions in your survey, it takes about one or two minutes per book to conduct the survey *after* the surveyor has enough experience to develop a familiarity with the process and to establish a good workflow. If you are working in the stacks using a handheld computer, it can save a great deal of time to have one person survey the book and answer the questions while the other person records the data into the handheld. After a while, both surveyors will memorize the questions and their order so that the person surveying the book will be able to go through the volume and answer the questions as fast as the other person can record the data.

A library can only have confidence in the results of the survey if they can have confidence that the data was gathered and analyzed properly. Ensuring the data is gathered properly means having a well-defined survey, with detailed instructions on how to gather the sampled items for the survey and how to answer the questions from the survey instrument for each item. Despite the best-laid plans, there will be times when volumes will be surveyed that will cause problems to the surveying team. Again, many of the questions are subjective. For this reason, it is important to have a backup plan where the questionable item can be set aside for review by other surveyors on the team so a consensus can be reached.

Another extremely important activity is to pretest the survey instrument and procedures. This can be done on a small sample of about fifty items. This will help you work the bugs out of the questionnaire and the procedures. For example, when I conducted my first survey using the instrument described here, for the question, "Number of circulations in the last 10 years" I had the first answer as 0 to 5. Only after a pretest did I realize that I needed to have an answer of zero and then a separate range of 1 to 5.

During the pretest have the surveyors keep a careful written record of the procedures they used in each step of the survey process. This seems like a lot of extra, unnecessary effort. I have often had survey teams complain about this step when I tell them about it, but after they have taken the time to take careful notes, they come to appreciate how this step helps them in developing a good surveying pattern. Also, recording small problems that seem insignificant at the time might later prove to be big issues that could make the results of the survey much less valid if they were not addressed at the beginning of the survey effort.

After the survey is created and the sampling method developed, and after a statistics expert has reviewed both for validity, then the survey can be conducted. At that point, the library has a powerful set of raw data, analysis of which can help the library in making preservation, collection development, and public services decisions for years to come. This leads us to our next chapter.

Notes

1. Derek Rowntree, *Statistics without Tears: A Primer for Non-mathematicians* (New York: Scribner's, 1981).
2. Arthur W. Hafner, *Descriptive Statistical Techniques for Librarians*, 2nd Ed. (Chicago: American Library Association, 1998).

Chapter 4

Analyzing the Data

After the data are collected, you will learn just how powerful database software can be in helping you to analyze your collections. Not only will you be able to successfully predict such things as what percentage of the collection circulates regularly, or what percentage needs repair, but you will also be able to combine data from questions to learn, for example, what percentage of paperback volumes have cover stiffeners placed on them and how well they have held up over time. You will be able to learn if dust jacket covers protect the books or if they actually cause the joint area to fail more quickly.

With the data recorded from a survey the important work begins. The analysis involves statistical expertise. Inappropriately analyzed data can actually be worse than having no assessment information at all. The classic mistake with statistical analysis is to try to gain more information from the findings than the data can support. After the data are gathered, a statistical expert can readily analyze the data or assist you in analysis to ensure proper results are obtained and that those results are properly interpreted.

For the most part, the information collected in library condition surveys is nominal data, meaning the data have no numeric value. The data are similar to a list of names. This means you can get counts on the data and percentages of those counts, but there is little ability for more sophisticated types of analysis. However, in general, counts and percentages is all the analysis most libraries will need. The advantage to this is it makes the statistics simple and straightforward to analyze.

When analyzing the data, it is a good idea to start by creating a large table with the results for every question from every library all on one table. This summary provides a good snapshot of the collections and will help you identify where further analysis is needed. Table 4.1 is an example of how this can be done. This table is from a series of conditions surveys my colleague, Bradley L. Schaffner, and I conducted in libraries in Ukraine and Bulgaria.

Table 4.1 enabled us to immediately identify what conditions were the same in all of the libraries surveyed, and which were different.

Where things were different we were able to do further analysis to compare and contrast conditions.

	L'viv, Ukraine				
	University	Academy	Combined	New University Building	Bulgarian National Library
Size of Volume					
Regular	95.37%	94.79%	95.11%	100.00%	97.78%
Folio	3.62%	5.21%	4.33%	0.00%	1.77%
Oversized	1.01%	0.00%	0.56%	0.00%	0.44%
Type of Volume					
Monograph	60.97%	70.97%	65.44%	97.59%	67.85%
Part of Multi-volume Set	25.55%	15.63%	21.11%	2.41%	11.75%
Serial	13.48%	13.40%	13.44%	0.00%	20.40%
Score	0.00%	0.00%	0.00%	0.00%	0.00%
Leaf Attachment					
Sewn Through the Fold	10.66%	8.93%	9.89%	8.43%	36.81%
Oversewn	0.60%	0.00%	0.33%	0.00%	0.67%
Adhesive Bound	64.79%	53.35%	59.67%	60.24%	49.00%
Stapled Through the Fold	11.07%	20.84%	16.56%	10.84%	10.42%
Side Sewn or Stapled	12.68%	14.39%	13.44%	20.48%	3.10%
Spiral or other Loose Sheet Bindin	0.00%	0.00%	0.00%	0.00%	0.00%
Condition of Text Block					
In Good Condition	80.68%	84.62%	82.44%	81.93%	84.67%
Remain in Stacks	11.27%	5.96%	8.89%	7.23%	7.11%
Needs Treatment	8.05%	9.43%	8.67%	10.84%	8.22%
Paper pH (Abbey Pen)					
Yellow or Clear (Acidic)	93.96%	95.02%	94.44%	96.39%	89.14%
Tan (Slightly Acidic)	3.82%	2.24%	3.11%	3.61%	1.55%
Purple (Alkaline)	2.21%	2.74%	2.45%	0.00%	9.31%
Paper Fold Test (Paper Breaks After)					
Less than 1 Fold	1.81%	0.25%	1.11%	1.20%	11.75%
Less than 1 Double-fold	8.25%	3.72%	6.22%	10.84%	3.10%
Less than 2 Double-folds	11.67%	5.96%	9.11%	20.48%	5.54%
Less than 3 Double-folds	4.63%	1.74%	3.33%	2.41%	3.55%
More than 3 Double-folds	73.64%	88.34%	80.22%	65.06%	76.05%
Brittle	21.73%	9.93%	16.44%	32.53%	20.40%
Paper Condition					
White and Strong	35.69%	31.51%	33.82%	26.51%	45.23%
Yellowish or Tan	51.41%	48.64%	50.17%	54.22%	37.03%
Brown	12.90%	19.85%	16.02%	19.28%	17.74%

Table 4.1: Results of Condition Surveys

One useful test for comparing results from different libraries such as in this case is called the chi-square test (χ^2). This test is designed to compare observed results to expected results with the null hypothesis being that there is no difference between the observed and expected results. In other words, the expected results are that if the same survey tools and methods are used in libraries in both Ukraine and Bulgaria, then the same type of results are expected. The chi-square test will let you know if the difference in results is statistically significant. For example, I used this test to evaluate the differences between the results I got from a condition survey I did in a library's stacks, and the results I got from surveying volumes returning from circulation. In many ways the collections were nearly identical, but in other ways they were significantly different. For example, materials returning from circulation were much more heavily used, were much newer, had been mutilated more, and required more repair.

The chi-square test is not complicated. In fact, many spreadsheet programs will run a chi-square test for you if you simply identify the two columns or rows of data you want to compare. But again, it is still good to check with a statistical expert to make sure you are applying the test correctly, and to make sure you are interpreting the results properly.

In addition to an overall table of results, there are certain results that nearly every library will want to look at. Some of these include paper condition, binding condition, overall condition of the volume, usage patterns, shelving condition, place of publication and/or language, and preservation needs. Other specific questions will be directed by the strategic initiatives of your library. For example, a library may want to determine if their preshelving treatment funds are cost effective.

We will look at why analyzing these results is important, but first it is important to discuss how to a use a database program to conduct the analysis. The easiest way to pull data from the database is to use queries. Some database programs have various macros and wizards created that will allow you to produce reports without setting up a query. These interfaces are getting more and more sophisticated, but for now, it is probably still best to set up your own queries to get to the data you want. Creating a query is fairly easy to do, and with a little practice you will find you can do so quickly. After creating the queries to get the results needed to create the table of complete results, you should be experienced enough to begin creating more sophisticated queries that pull results from multiple questions. For example, you will be able to query the data to find out what percentage of paperback volumes have received paperback cover stiffening treatments. Or, what percentage of U.S. imprints from the past ten years is printed on acid-free paper.

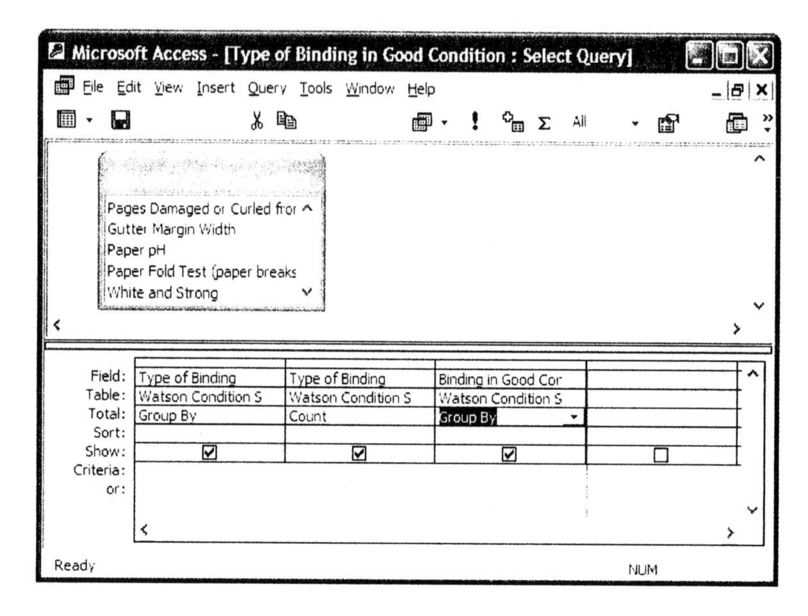

Figure 4.1: Create Access Query

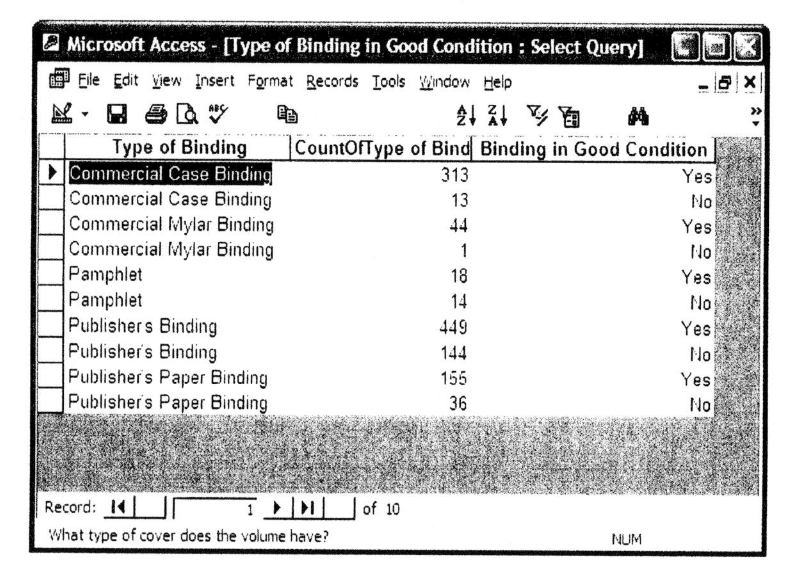

Type of Binding	CountOfType of Bind	Binding in Good Condition
Commercial Case Binding	313	Yes
Commercial Case Binding	13	No
Commercial Mylar Binding	44	Yes
Commercial Mylar Binding	1	No
Pamphlet	18	Yes
Pamphlet	14	No
Publishers Binding	449	Yes
Publisher's Binding	144	No
Publisher's Paper Binding	155	Yes
Publisher's Paper Binding	36	No

Figure 4.2: Access Query Results

Once you are used to it, creating the queries is easy, but analyzing the results of the query can be more challenging. Again, most database programs have excellent report writers that will produce nice reports from the queries you create. However, it is often faster and easier to pull the results into a spreadsheet program that provides you with great number crunching and sorting abilities. Most office software suite packages make converting query results into a spreadsheet table a very seamless process. Or you can simply cut the table from the database program and paste it into the spreadsheet program.

Figures 4.1 through 4.4 show how you can go from creating a query to having meaningful results about your library's collections. The following query (figure 4.3) is to identify, by percentage, the type of bindings on volumes in the University of Kansas's Watson Library, which is the main library on campus. The query also identifies what percentage of each binding type is in good condition.

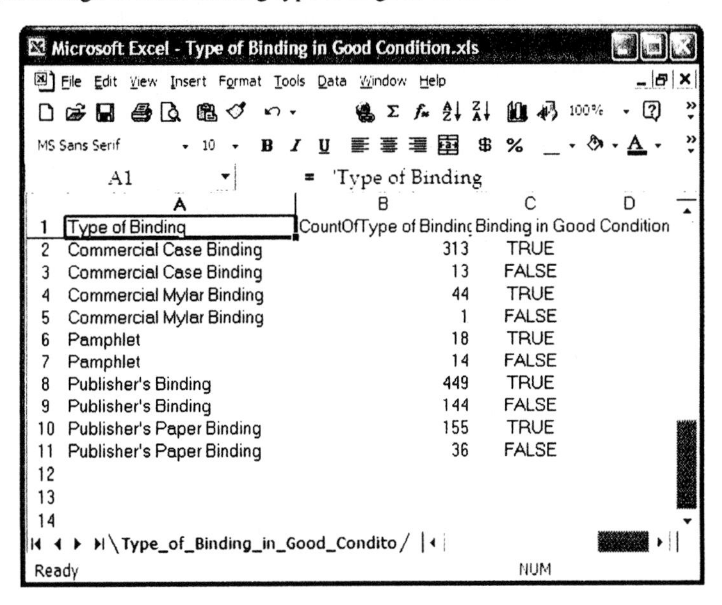
Microsoft Excel - Type of Binding in Good Condition.xls

	A	B	C	D
1	Type of Binding	CountOfType of Bindinc	Binding in Good Condition	
2	Commercial Case Binding	313	TRUE	
3	Commercial Case Binding	13	FALSE	
4	Commercial Mylar Binding	44	TRUE	
5	Commercial Mylar Binding	1	FALSE	
6	Pamphlet	18	TRUE	
7	Pamphlet	14	FALSE	
8	Publisher's Binding	449	TRUE	
9	Publisher's Binding	144	FALSE	
10	Publisher's Paper Binding	155	TRUE	
11	Publisher's Paper Binding	36	FALSE	
12				
13				
14				

Type_of_Binding_in_Good_Condito

Figure 4.3: Query Results Transferred to Excel

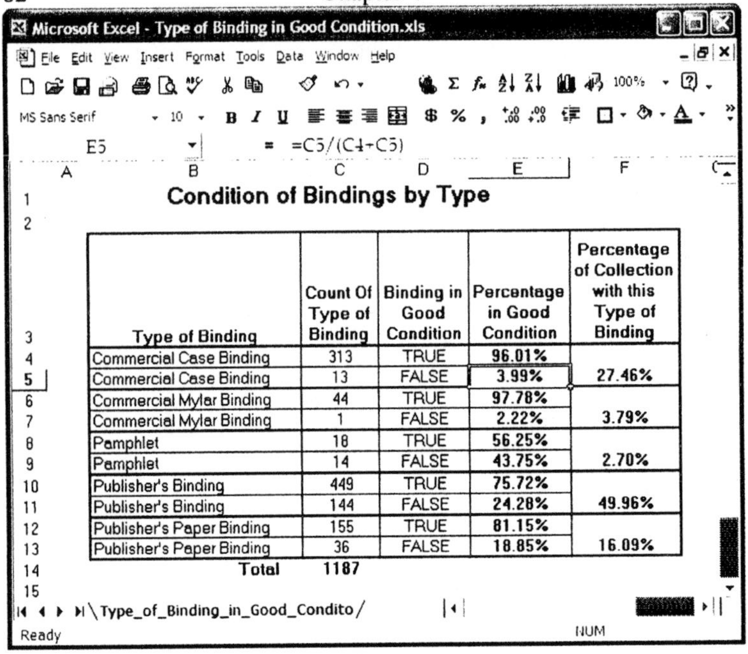

Microsoft Excel - Type of Binding in Good Condition.xls

E5 = =C5/(C4+C5)

Condition of Bindings by Type

Type of Binding	Count Of Type of Binding	Binding in Good Condition	Percentage in Good Condition	Percentage of Collection with this Type of Binding
Commercial Case Binding	313	TRUE	96.01%	
Commercial Case Binding	13	FALSE	3.99%	27.46%
Commercial Mylar Binding	44	TRUE	97.78%	
Commercial Mylar Binding	1	FALSE	2.22%	3.79%
Pamphlet	18	TRUE	56.25%	
Pamphlet	14	FALSE	43.75%	2.70%
Publisher's Binding	449	TRUE	75.72%	
Publisher's Binding	144	FALSE	24.28%	49.96%
Publisher's Paper Binding	155	TRUE	81.15%	
Publisher's Paper Binding	36	FALSE	18.85%	16.09%
Total	1187			

Type_of_Binding_in_Good_Condito

Ready NUM

Figure 4.4: Example of Reporting Table

This one example demonstrates a great deal. It shows how easy it is to set up a query (figure 4.1) that pulls data from more than one question in the survey. It shows how detailed information about a library's collection can be pulled from the survey results (figure 4.2), and it shows how sometimes it is easier to manipulate the data using a spreadsheet program rather than relying solely on the reporting functions of the database program (figure 4.4).

The results of this one query also reveal a great deal about the make up of the Watson Library collections. It shows that nearly half of the collection has publisher bindings, but that one in four of those volumes are not in good condition (figure 4.4). In fact, the publisher paper bindings are in slightly better condition than the hardbound books. However, this is where statistics can be deceiving if you don't know the history of the collection. In Watson, like many academic libraries, there was a concerted effort to bind as many paperback bound volumes as possible. For this reason, paperbacks make up only 16 percent of the collection. However, in recent years funds were not available to bind as many paperbacks, so they had to go to the stacks without binding. To validate this, another query was conducted that showed that a much higher percentage of paperback volumes had relatively recent imprint dates compared to the hardbound volumes. Still, based on the results of this survey, it was decided at the University of Kansas Libraries that we would purchase more materials with paperback bindings to save collection development funds, and that we would be more selective in what paperback volumes get bound since many kinds of paperback volumes hold up well in research library collections.

Another important piece of information that comes from this query is that commercially bound volumes hold up extremely well, with over 96 percent of all commercially bound volumes being in good condition (figure 4.4). This demonstrates just how effective library binding can be as a preservation treatment when it is used properly.

At the University of Kansas Libraries, we make an effort not to send pamphlet bound volumes to the stacks without some kind of preshelving treatment. However, through the years enough unbound pamphlets have made it into the stacks to make up about three percent of the collections. The findings validate our efforts to pretreat pamphlets because nearly half of the unbound pamphlets in the Watson Library stacks are not in good condition.

With this brief demonstration of what information is possible to gather from the condition survey, let us turn our attention to the areas that every library will want to look at.

Paper Condition:

The majority of library collections are still paper-based materials. The importance of the condition of the paper in those volumes cannot be understated since paper condition is the primary factor in determining how long a volume will be useable, and what preservation treatments can be performed on the volume. Between the skills of a commercial bindery and a well-trained book repair or conservation professional, most volumes can be successfully, and often economically, rebound—regardless of the condition of the original binding. However, if the paper in the volume is brittle, it will not have the strength to withstand rebinding or future use. For a volume with brittle paper, the only choices are to withdraw the original, replace it with a new copy, or reformat it. These are labor-intensive, expensive processes.

Residual acids in the paper left during the manufacturing process causes paper to deteriorate. Over time, these acids react with oxygen and moisture in the air to break down the paper fibers, causing the paper to become weak and brittle. This chemical reaction is greatly accelerated by high temperature and high relative humidity and by rapid fluctuations in temperature and relative humidity.

A lot of national attention has been given to the brittle paper problem. Many programs have been established in research libraries in an effort to reformat brittle volumes. These programs were begun as libraries and archives became aware that up to 25 percent of their collections had brittle paper.

There have been a lot of conditions surveys conducted in large research libraries throughout the country, but you cannot simply assume that your brittle paper rates will be similar to other survey results you have seen. A stable environment for books can make a major difference. So too can geographic location. At the University of Kansas, we found only 6 percent of our holdings have brittle paper, and libraries in the intermountain west with relatively cool dry summers and well air-conditioned libraries have an even lower percentage of brittle holdings. Therefore, it is important to know the condition of the paper in the collections in your library.

Just knowing the rate of brittle paper is not enough. Many libraries find they have large numbers of brittle volumes in their stacks, but that a relatively small percentage of those volumes circulate regularly. Use the circulation data to see what percentage of your brittle volumes regularly circulates. This will help you project what the real needs are for the collection. For example, at the University of Kansas, while 6 percent of the volumes in the stacks have brittle paper, less than 2 percent of the volumes that circulate have brittle paper. Still, the number of brittle volumes returning from circulation that need treatment before they can be used again represents several hundred volumes per year. At $80 to $100 per volume, 200 brittle volumes can cost nearly $20,000 to reformat. This helped us decide what funding levels we should establish for our reformatting efforts.

In addition to knowing what percentage of the collection is brittle today, it is important to predict what will happen to the collection in the future. For example, what percentage of your collection is printed on acidic paper? Thankfully, most publishers in the United States and Western Europe now print nearly all of their books on acid-free paper—especially hardbound volumes and scholarly paperbacks. This means that your library will eventually be past the brittle book problem. However, it will remain a serious concern for the next century. At the University of Kansas, over 75 percent of the collect is printed on acidic paper that will eventually become brittle. Fortunately, only 50 percent of the volumes returning from circulation were printed on acidic paper, showing that there is light at the end of the tunnel.

Some libraries face a much stiffer challenge. In the libraries Brad Schaffner and I surveyed in Eastern Europe, 20 percent or less of the volumes had brittle paper, but over 90 percent of the volumes in the collections were printed on very acidic paper that will, due to poor environmental conditions, become brittle in the next ten to twenty years.

Knowing the condition of the paper in the volumes in your collections will help your library effectively plan for the future. A research library that wants to keep its collections in tack will need to begin making plans for reformatting and replacing brittle volumes at a steady rate, year by year.

Text Block Condition:

A bound volume is made up of two parts, the text block, or the bound pages, and the cover. Data from the survey relating to each part will be looked at in turn. These two components were considered separately when collecting data for the survey because the physical integrity of volumes will fail for different reasons depending on how their text blocks are put together and how their covers are made.

The primary component of a text block is paper, which has been discussed in detail. Next, it is important to consider how the pages of the text block are held together. This is called leaf attachment. Knowing how the various kinds of leaf attachment methods hold up to use in your library will shape binding and prebinding efforts.

It is important to look beyond just the overall condition of the text block. Compare the gross findings with results of a finer analysis. For example, how well do the various leaf attachment methods and other features of the text block hold up for volumes that are heavily used? At the University of Kansas, we found that most leaf attachment methods hold up pretty well with the only real exceptions being spiral bound and other loose bound text blocks like three ring binders and such. Side sewn or stapled volumes were also inferior. However, when we compared how well leaf attachment methods stood up to heavy use, we found that volumes that were sewn through the fold lasted much better than other methods.

As expected, we also found that double-fan adhesive bound text blocks performed by commercial binders using high quality, cold polyvinyl acetate (PVA) adhesives were much superior to the hot melt, quick drying adhesives used in many publisher bindings. These hot melt adhesives are often stiff and brittle, and lack the adhesion properties of a good PVA.

Binding Condition:
The type of library, and the amount of use materials in the collection receive, impact how well various binding styles will hold up in your library. For example, in large research libraries it was found that paper-back volumes hold up pretty well and may not need prebinding. This is not as true for public libraries or heavily used collections. However, until your library looks at the data that is specific to your library, you will not know what kind of bindings hold up well in your libraries, and which ones do not.

Binding condition is especially important if you have materials in your collections that are easily obtainable in more than one binding style such as paperbacks versus hardbound volumes, or library bound children's books versus publisher-binding children's books. You may find that the text blocks are the weak link for children's books, and that the pages are torn beyond repair far sooner than the covers are dam-aged. If this is the case, it may not be worth the extra cost of purchasing children's books with library bindings.

Overall Condition of the Volume:
For preservation purposes, the last question on the survey is extremely important. It takes into account the answers from the other questions about the text block, and the covers and the paper quality, and makes an overall assessment. This question needs to reflect either the way your library currently handles damaged materials, or it should reflect the goals the library has for future preservation efforts. For example, if your library has a well-developed book repair program the answers about how to treat the overall condition of the volume should take those services into account. Likewise, if your library plans to build a conservation lab, then the answers to this question should include the category of items needing conservation treatment. This will help your library plan for how large of a conservation program is needed to handle damaged items from the general collections.

The answers to this question will let the library know how many items in the collections need preservation treatment, but these figures can be somewhat misleading and should be interpreted properly. For example, many large, older research libraries are going to have collec-tions that are made up of older materials that do not receive very much use. A high percentage of these older materials are going to be in need of some kind of preservation treatment, but if your library selects items for treatment based on usage like most libraries do, then having a lot of damaged books on the shelves that will rarely get used does not represent a preservation nightmare.

At the University of Kansas, we used the same survey instrument and procedures that we used to survey the general collection stacks to also survey materials returning from circulation. This provided a lot of very useful data and we saw that in many important ways, the materials returning from circulation are not representative of materials in the general collections. One of the most useful pieces of information from the circulation return survey was the overall condition of the materials. This data allowed us to draw very accurate predictions about how many volumes returning from circulation needed preservation treatment. Furthermore, it enabled us to predict the number of volumes per year that would need book repair, conservation treatment, commercial binding, brittle book processing, or to be placed in an enclosure. Real data of what the preservation department has treated over the years has proven to be remarkably consistent with the predictions we made based on our condition survey results.

Usage Patterns:
How you set up your survey in terms of gathering usage data will determine the conclusions you can draw about general circulation patterns for your collections. However, some of the cross analysis you can do is to compare how the language of the publication impacts the amount of use it gets, or how the physical format impacts use. For example, it is commonly believed that volumes in public libraries will not circulate as well if they have a sterile, utilitarian looking library binding. Your survey results will allow you to see if library bound volumes circulate less than books with publisher bindings.

Another useful piece of information is to see how long it takes for a volume to circulate after it is added to the collection. A library can spend a great deal of time and effort trying to rush materials through the processing stage and onto the shelf where it is available to patrons. However, if the patrons do not value this service it might be a waste of resources to hurriedly make materials available if they are not immediately used. This is an example of where a survey of the public about what services they want may produce different results than the collection survey. For example, a patron may fill out a survey asking them if it is important to have the books processed and made available as quickly as possible. Well, of course, everyone is going to say that such a service is important. They may even believe that such a service is very important, but by surveying the collections a library can find out, on average, how long it takes for an item to circulate after it is made available to the public. Statistical data, based on behavior patterns, is always going to be more valid that data based on opinion or self-reporting.

The usage pattern information is useful for making long-term preservation plans, but it is often most useful for collection development purposes. It helps identify what is being used and how often. If you are able to use the circulation system to pull the usage data, then you can identify what subject areas are receiving the most use, and how collections are used over time in your library.

Shelving Condition:
This data is important to know about the stacks. It evaluates the effectiveness of your shelf reading program, your shelving unit, and can provide insight into what the patrons are doing to your stacks. For example, at the University of Kansas Libraries we found that shelving problems were not universal throughout the collections. Some areas had problems with being overpacked, other areas did not. Areas of the collection that received a lot of patron browsing, such as popular fiction, had a higher percentage of items that were shelved in the wrong location. The literature section, with their long complicated call numbers, also had a higher percentage of misshelved books.

Some of the problems reported by this question mean that the one book surveyed has a problem. For example, if a book is shelved in the wrong location, it means that just that book might be wrong, but if a book is shelved too tightly, then it means all the books on that shelf are packed too tightly. This means the problem could be multiplied by twenty or more times.

It is important to report this data carefully. A high number of misshelved items does not necessarily mean that the shelving unit is not doing a good job. It may simply mean that the collection is highly browsed by patrons. However, regardless of whose fault it is, it probably means that your library needs to invest the time and effort to conduct a careful shelf reading.

Place of Publication and/or Language:
Knowing place of publication or the language that the volume is published in can provide insight into the makeup of your collections and how they are used. This can shape the collection development activities of your library. It is probably not a wise use of resources to purchase a lot of materials in Chinese if your library does not support a Chinese population. For many languages, this is obvious without conducting a

survey, but for others you might be surprised at what you find. For example, some library collections I have surveyed showed that foreign language publications did not circulate as much as English language publications. This is what you would expect. However, what was not anticipated was that the foreign language materials received much more use than expected via interlibrary loan. If your library has reciprocal agreements for interlibrary loan, this might be important information.

The data gathered about the place of publication and publication language is very useful for helping libraries identify library use patterns for area studies units on campus and can help direct collection development procedures in those areas. The data on language and place of publication helps identify a subset of materials used by specific area studies. It is important not to overanalyze the data because subset data can quickly lose its statistical significance, but it can provide some snapshot information. For example, in one survey I conducted I found that Spanish language books written about environmental issues received higher than anticipated use. This data shaped the purchasing decisions of the bibliographers of Spanish and Environmental Sciences.

Preservation Needs:
One of the most useful aspects of a condition survey is the wealth of information it can provide about the preservation needs of the library's collections. You can quickly identify what percentage of materials has been bound by a library binder, what percentage needs to be bound, what percentage of the collection has brittle paper, and so forth.

At the University of Kansas Libraries, we were able to determine that paperback volumes hold up on the shelves almost as well as hardbound volumes do, so we were able to greatly reduce the amount of prebinding we did. This allowed us to shift some of those funds towards other preservation activities such as reformatting brittle books.

One of the very best preservation uses of condition survey data is to determine what percentage of the collection has been repaired, and how well those repairs are holding up over time and after use. Some repairs hold up very well to repeated heavy use. Some repairs seem to be a good idea when they are first administered to the book, but then, over time they breakdown along inherent weaknesses unwittingly introduced into the repairs. For example, one library I surveyed had made use of a large number of premade binders. These binders seemed sturdy and provided good structural support to flimsy items. However, over time the spine piece on the binders began to deteriorate and tear. The result was that the library had thousands and thousands of binders to replace. The survey helped identify the problem, gave numbers on how severe the problem was, and made it possible for the library to create a policy to pull for rebinding all items that returned from circulation in one of these ineffective premade binders.

Useful as the preservation data is, it is important to not overanalyze it. For example, it is a common mistake to assume that if a survey shows that about 20 percent of the volumes in the collection need some kind of repair that the library needs to increase funding to address this problem. That is not necessarily the case because the makeup of materials in the stacks do not always mirror the makeup of the materials that circulate. A survey of the stacks may show that 20 percent of the volumes need repair, and only 10 percent need commercial binding. However, a survey of materials returning from circulation will probably show very different results, and since most libraries primarily select materials to send for preservation treatment by identifying damage materials as they return from circulation, the stacks population is not necessarily the data you want to use for making specific long-term preservation decisions.

There are a few common themes that people look at when they analyze data from a condition survey, but don't limit yourself. Be creative. The survey instrument and analysis tools described above are extremely powerful and flexible. A wealth of information about your library and its collections can be gleaned from the data gathered. Do not underestimate the power of the data you gathered.

It is also important to keep the data alive. Make sure the data is migrated forward to new versions of the database software program you are using, and refer to the data regularly to answer questions that might come up later about the collection.

The condition survey is not a task that comes to a neat end with a completed report. The data remains alive and adaptable, and so too should the analysis. Even a very thoroughly conducted survey will not answer all the questions a library has about its collections. Instead, it will point out new questions and call for new assessment strategies. That leads us to our next chapter.

Chapter 5

Assessment and Planning

Assessment has become an important part of library activity. We have to assess what services we offer and informational resources we provide on an ongoing basis because of the rapid technological changes that are taking place in information industries. Libraries are constantly faced with inflating serial prices, increasing electronic sources of information, and limited resources. They need assessment information to assist in making purchasing decisions, and to evaluate how effective past decisions were in meeting the library's goals.

By now, you have come to realize that this book offers no magic spells or silver bullets. It does not contain a road map to the oracle of Delphi. It simply describes some assessment tools and strategies that libraries can utilize help inform their decisions. When a library takes the time and effort to conduct a thorough collections assessment, it will gain information that will help it better serve its clientele, but it will not make the decisions completely painless. A good assessment will often introduce as many questions as it does answers. That is why it is important to use assessment tools that are dynamic and produce data that can be analyzed in a number of ways to generate the kind of information a library needs to help inform its decisions.

This book has introduced a pragmatic assessment method a library can utilize to gain a wealth of information about its collections and how those collections are used. This method is based on sound assessment strategies and statistical rigor to ensure that the data gathered truly represent the collection as a whole and provide meaningful information to direct library activities. This tool is adaptable so that each library can alter it to gather the information about their collections the library needs most. The process of conducting the assessment is demanding, but not overly time consuming. The goal was not to create a morass of assessment reports and tools that would be overburdening to a library.

This collection assessment method should prove very useful to the collection development and preservation efforts in any library that uses it. It should also be extremely beneficial in helping direct future assessment efforts. For example, perhaps a public library finds that very few of their library bound children's books circulate. This information can help shape a user survey. Instead of the question, "Do you or your children choose library bound volumes from the shelf when browsing the collections?" that will, at best, validate the data you have already gathered from your condition survey, you can ask a question such as, "It is clear that library bound children's books do not circulate very often. What, if anything, do you or your children find unappealing about library bound children's books?" You can leave the question open ended, or give them choices or answers to choose from if you don't want open-ended questions. The difference is that your library will end up gathering useful data about why a class of material is not used very much instead of just confirming what you already know.

Assessment is a continuous and ever-changing activity. The collection assessment tool you develop for your library will probably need to be altered before using it again because you will look at the data gathered and wonder about other information you wish you had gathered.

Always remember, that the primary purpose of surveying a collection is to provide a library with the data needed to make informed decisions. When you get ready to conduct your survey, there will be many suggestions from staff on what data should be gathered, but remember to be practical and do an assessment that is relatively easy to conduct and produces the data that is of most value to your library. One of the guiding principles behind the need for assessment is the law of diminishing returns. Businesses have always had to pay close attention to the law of diminishing returns if they want to be successful. Libraries are relatively new to these economic models of efficiency. Libraries need to know, for example, how long, on average, does a book sit in the stacks after it returns from circulation before it circulates again. If it is a long time, then maybe the library can save staff resources by not re-shelving items as fast as it used to. Or, it may find that it needs to re-shelve things more quickly than it has been doing, and by doing so they reduce the need for as many copies of a title.

When conducting an assessment, a library must remember the law of diminishing returns. In general, a library will want more information than it can afford to gather. It is very easy to suggest the creation of a myriad of reports and assessment tools without taking into account the staff time and energy needed to create these tools, or to give full consideration to how the data acquired would be used. Often the data requested, while interesting, may only be tangentially related to the main mission of the library.

Make sure the people doing the assessment understand how the collection assessment tool works, and the statistical principles behind how the data is gathered and analyzed. Otherwise, the effort will produce erroneous results, and misused data can contribute to poor library decisions. It is important to know how to use and interpret the data correctly and to come to conclusions that are supported by the data. This can be challenging at first, and that is why it is important to secure the assistance of statistical experts until library staff are more comfortable with the process.

Libraries and librarians can use the data produced by the collection survey a number of ways. For example, a new bibliographer could learn a great deal about the collections she is responsible for. The survey data can tell her about the makeup of the collections, how the collections were developed, and how they are being used.

In the rapidly changing information age we live in, libraries have to be adroit at knowing the informational needs of their clientele, and knowing what information exists that can fill those needs. Being on top of that game today is no guarantee that your library will be there tomorrow. The only way to guarantee that your library is moving in the direction it needs to go is through constant assessment and careful planning. Assess the information resources available. Assess the needs of the patrons. Assess how the information that your library has acquired is being used. This information will help guide future decisions.

Three examples will demonstrate how assessment information can help your library with their planning efforts. The first example shows how data from one library can be used for planning in other libraries. I helped a library conduct a survey that had been placing barcodes on the outside of their books for many years. The University of Kansas Libraries were thinking of starting this practice. Therefore, while we were conducting the survey of the other library's collections we included a question to see how well their barcodes were performing on the outside of their books. We looked to see if the barcodes were peeling off, or if the adhesive was oozing, or if they were becoming abraded or soiled to the point that they did not work any longer. Because we were scanning the barcodes of each sampled book into the database, we could readily test if they were still readable. Based on the results of this question from the other library's survey, we learned that barcodes seem to hold up well on the outside of research library books. Therefore, we could, with greater confidence, start placing barcodes on the outside of the volumes at the University of Kansas Libraries.

Another example shows how data can be used years later to help make tough preservation treatment decisions that were not imagined at the time the survey was conducted. As the anthropology bibliographer at the University of Kansas Libraries, I conducted a thorough survey of the anthropology collections in the KU collections to determine how they were being used and what kind of subject areas were covered through the last hundred years of purchases. There were some preservation related questions in the survey, but mostly the questions focused on collection development issues. A couple of years later a water pipe burst in the main library sending a flood of water cascading over some large, old folio archeology volumes. We dried them quickly, and most of them were able to return to the stacks, but there were several dozen large, old volumes that were in need of preservation treatment. We were left with a decision on whether to rebind these volumes, or to do minimal repair on them, or to box them. I decided to look at the usage patterns for these materials that I readily had for these materials from the anthropology survey I conducted. After a few queries I learned that these materials did not receive heavy use ever, but that they received constant steady use. The data made it clear that the materials would be regularly used in the future. This made it worth the time and effort to rebind these large awkward volumes in our conservation unit.

The final example is from some surveys Brad Schaffner and I conducted in Eastern Europe. We found that most of the library collections we surveyed did not have nearly as high of a brittle book problem as we had expected to find. We learned that this was because most of the books were stored in large, old stone buildings with thick walls that moderated the temperature and relative humidity extremely effectively. As a result of these nearly ideal environmental conditions, the brittle book problem was relatively small in most of the libraries, but there was one newer building that we surveyed that was built in the 1980s. This new building was overheated in the winter and not cooled in summer. As a result it was nearly always hot and sometimes quite humid. This caused the books to cook and accelerated their decomposition. In less than twenty years in the new building the brittle rate more than doubled. This was proof positive of the beneficial effect that can be gained through having good environmental conditions.

Many of these collections were stored in large old stone churches that had been taken over during the Soviet period to store books. With religious freedoms being restored, the churches were being returned to the people. This meant the libraries had to get their collections out. They were, at the time of our surveys, looking for new locations. Unfortunately, most of the buildings available to the libraries were like the new library building with poor environmental conditions. As a result, if they move their collections into these newer buildings the collections will become extremely embrittled in less than 25 years. Because of the extremely strong evidence we gathered we were able to produce a report for these libraries explaining the need to move their collections into environmentally sound buildings. In the end, it will be for the libraries to decide, and circumstances may prevent them from establishing an ideal environment, but at least they have the information they need to make a strong strategic case for gaining the best environment for their collections possible.

We were also able to use this data back here in the United States to demonstrate to librarians at national conferences just how important it is to have good environmental conditions. We were also able to use this data to help us secure a remote storage facility with better environmental conditions than we otherwise would have had because we were able to convince library and university administrators of the importance of temperature and relative humidity controls.

Assessment data can be used to help with strategic planning and to make day-to-day decisions. At the University of Kansas Libraries, we used the results of our condition survey to decide on strategic initiatives. For example, when we saw that paperback volumes held up well in our stacks, we switched to a paper preferred acquisitions policy in which we have standing order with our book supply vendors to send us the paperback volume and not the hardbound volume if the publisher produces both at the same time. This was a strategic decision. On day-to-day matters, we found that the spine repair treatment we were using was not holding up very well to use. The heavy cloth we were using in the repair was causing stress on the inner joints, causing them to fail on some books. We were able to use this data to make adjustments to the treatment to make it better.

There are lots of examples of how to use the assessment data to improve your library's services. Many of the results will be useful immediately in your planning efforts. Other results will prove useful often years later as the library faces unforeseen problems for which the survey data can help with the decision-making process.

The bottom line is to do the survey in a consistent, thorough manner so that the data will be valid and meaningful in assessing the collections. Analyze the data appropriately so that the results will accurately predict the condition of the collections. And carefully document the procedures used to conduct the survey and data analysis so that the data can be used in the future, or so that the survey can be repeated in the future.

Assessment is an important activity in any library—especially in these times when information technology is changing so rapidly and patrons' informational needs are more complex than ever. As important as assessment is, it is important to make sure that it is done properly. Make sure that the assessment effort does not cost too much time and effort. The assessment methods discussed in this book are carefully designed not to take too much time or cost too much. Through experience with assessment activities a library will learn to conduct assessment activities more efficiently, and to gain more useful information from each assessment activity. But to gain that know-how you must begin.

Bibliography

Baird, Brian J. "Paperbacks vs. Hardbacks: Answers from the University of Kansas Libraries' Condition Survey." *Abbey Newsletter* 20, No. 6 (December 15, 1996): 93-95.

———. *Preservation Strategies for Small Academic and Public Libraries*. Lanham, MD: The Scarecrow Press, 2003.

———. "Those Amazing Library Bindings." *New Library Scene* 15 (October 1996): 9-10.

———. *University of Kansas Libraries' Preservation Department Web Page* at: http://www2.lib.ku.edu/preservation.

Baird, Brian J., Jana Krentz, and Brad Schaffner. "Findings from the Condition Surveys Conducted by the University of Kansas Libraries." *College & Research Libraries* 58 (March 1997): 115-126.

Baird, Brian J., and Bradley L. Schaffner. "Extinguishing Slow Fires: Cooperative Preservation Efforts." *Racing towards Tomorrow*, Proceedings of the Ninth National Conference of the Association of College and Research Libraries April 8-11, 1999. Edited by Hugh A. Thompson. Chicago: Association of College and Research Libraries, A division of the American Libraries Association, 1999: 228-233.

———. "Slow Fires Still Burn: Results of a Preservation Assessment of Libraries in L'viv, Ukraine and Sofia, Bulgaria." *College & Research Libraries* 64, No. 4 (July 2003): 318-330.

Hafner, Arthur W. *Descriptive Statistical Techniques for Librarians*. 2nd Ed. Chicago: American Library Association, 1998.

Kaplan, Robert S., and David P. Norton. *Translating Strategy into Action: The Balanced Scorecard.* Boston: Harvard Business School Press, 1996.

———. "The Balanced Scorecard—Measures That Drive Performance." *Harvard Business Review* 70 (January-February 1992): 71-79.

Mead, Robert A., and Brian J. Baird. "Preservation Concerns for Law Libraries: Results from the Condition Survey of the University of

Kansas Law Library." *Law Library Journal* 95, No. 1 (winter 2003): 69-86.

Osif, Bonnie A., and Richard L. Harwood. "Statistics for Librarians." *Library Administration & Management* 15 (winter 2001): 50-55.

Patkus, Beth. *Assessing Preservation Needs: A Self-survey Guide*. Andover, MA.: Northeast Document Conservation Center, 2003.

Rowntree, Derek. *Statistics without Tears: A Primer for Non-mathematicians*. New York: Scribner's, 1981.

Schaffner, Bradley L., and Brian J. Baird. "Into the Dustbin of History? The Evaluation and Preservation of Slavic Materials." *College & Research Libraries* 60, No. 2 (March 1999): 144-151.

———. "Simple Steps for Evaluating the Condition of Slavic Collections: How to Determine if Your Collection Is at Risk." In *Libraries in the Age of the Internet*, ed. Herbert K. Achleitner and Alexander Dimchev, 217-226. Sofia, Bulgaria: Union of Librarians and Information Services Officers (ULISO), 2001.

Index

About the Author

Brian J. Baird is Director of Preservation Services at Heckman Bindery. Before that he was a faculty member at the University of Kansas for ten years as their first preservation librarian. He has also worked as a conservator at Princeton University Libraries. Brian holds a master's degree in library and information science from Brigham Young University. He also holds a bachelor's of science degree in psychology from Brigham Young University. Brian has been very active in the library preservation profession and has published a book and several articles on library preservation topics. He has assessed collections in more than a dozen libraries in America and Europe. Brian is married, and he and his wife, Jennifer, have seven children. To learn more about Brian and his professional activities, visit the Heckman Bindery web page at: http://www.heckmanbindery.com.